1424 FUN, RANDOM, & INTERESTING SPACE FACTS THAT EVERYONE NEEDS TO KNOW

Planets, Galaxies, Moons, and More

PROFESSOR SMART

A Thank You Gift

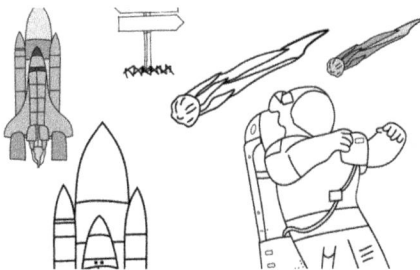

As a token of our appreciation, we would like to share our space coloring worksheet with you! To download this for free, visit bit.ly/space-coloring for more information.

All About Space

Space is cool!

And it's really cold!

And it's also really big!

Outer space starts at around 100 km (60 miles) above Earth. There is no air to breathe and it looks black to the human eye. However, space isn't empty ... it is full of gas, dust and other bits of matter.

Nobody knows how big space is and we measure distances in light-years. Telescopes allow us to see galaxies up to 13.7 billion light-years away and we are still unsure if our universe is the only universe that exists.

In this book, you will find 1424 fun, random, and interesting facts about moons, planets, the sun, galaxies, the ISS (International Space Station), asteroids, and more.

We hope you enjoy reading this book!

The Professor.

1424 Space Facts

1. The moon is getting further and further away from Earth by about 3.8 cm (1.5 in) every year.
2. Galileo had a refractor telescope that could magnify an image 30 times.
3. Trainee astronauts previously needed military jet test pilot and engineering experience. However, these days, high achieving engineering, science, or maths students can train to become astronauts.
4. The ISS weighs approximately 1 million pounds (453 tonnes).
5. An eclipse occurs when the moon, sun, and Earth all line up.
6. Miranda is one of Uranus' most exciting moons, with ice canyons, terraces, and other strange-looking surface areas. It looks like it has been turned inside out.
7. The sun moves at 220 km per sec (136 mi per sec).
8. Just like Jupiter, Saturn has oval-shaped storms.
9. BPM 37093, a star about 20 billion light-years from Earth, is a large diamond that weighs about 10 billion trillion trillion carats and is about the

size of the moon. It is also called 'Lucy' after the Beatles' song "Lucy in the Sky with Diamonds."

10. Inner planets are denser than outer planets, so they rotate slower than the outer planets.

11. Charon's surface, which is covered in ice, differs from Pluto's, which is covered in frozen nitrogen, methane, and CO2.

12. The greenhouse effect on Earth protects humans and moderates the Earth's global temperature. Without the greenhouse effect, the temperature would be -18 degrees C (0 degrees F) instead of 15 degrees C (59 degrees F).

13. If life is found on another planet, the Office of Planetary Protection department of NASA will deal with it.

14. Galaxies are classified into four main types by their shape - elliptical, spiral, lenticular, and irregular.

15. The same side of most moons always faces its planet. This is known as 'tidal lock.' Hyperion, Saturn's moon, is the only exception, and it rotates in different directions due to Titan's gravity.

16. NASA's Viking 1 and 2 both landed on Mars several months apart in 1976 and provided the first color pictures of Mars.

17. In 1966, the first successful uncrewed spacecraft landed on the moon. It was a Luna space probe belonging to the Soviet Union.

18. Venus is sometimes referred to as Earth's sister planet.

19. Titan orbits Saturn at a distance of about 1.2 million kilometers (750,000 miles) and takes 15 days and 22 hours to complete an entire orbit.

20. It takes 100,000 years light to reach Earth from

the sun's core but only 8 minutes 22 seconds from the sun's surface.

21. There are three types of satellites used for different purposes. Satellites for communication such as voice, data, and video transmissions, are fixed satellites. This makes up most of the satellites that are sent to space. Some satellites are made for GPS and navigation and positioning. In contrast, many others are used for scientific research such as space observations, earth science, and meteorological data.

22. You can see Saturn without a telescope.

23. The US and Russian astronauts keep separate water supplies on the ISS.

24. Venus is covered with sulphuric acid clouds, which means that it was not possible to view its surface until radio mapping, developed in the 1960s, observed its extreme temperatures and hostile environment.

25. In 2013, a huge meteoroid fireball entered Earth's atmosphere above Chelyabinsk, Russia, at a speed of 18 km per sec (11 miles per sec). It exploded 23 km (14 mi) above the surface and generated a shockwave that injured 1,600 people.

26. Mars' gravity is weak, so it doesn't hold onto its atmosphere very well.

27. Europa, Jupiter's moon, is the smoothest object in our Solar System due to its icy surface with no mountains and only a few craters.

28. The biggest reflecting telescope used for observing space is the Hubble Space Telescope.

29. 2002 AA29 is one of Earth's co-orbital satellites with a diameter of 60 meters (196 feet) and makes a horseshoe-shaped orbit around Earth. It gets

closer to our planet every 95 years and maybe a good space mission in the future.

30. The brightest galaxy in the M51 group is the Whirlpool Galaxy.

31. In a total solar eclipse, we can only see the solar corona of the sun.

32. The moon surface has been bombarded by micrometeorites over time, covering it in a layer of crushed powdered rocks and dust.

33. More than two-thirds of all galaxies are spiral galaxies. A spiral galaxy has a flat, spinning disk with a central bulge surrounded by spiral arms. As it spins at speeds as fast as hundreds of kilometers a second, it causes matter in the disk to take on a spiral shape, like a cosmic pinwheel. The Milky Way is an example of a spiral galaxy.

34. The Catholic Church persecuted Galileo for centuries as they thought he was blasphemous when he suggested the Earth orbited the sun. Pope John Paul II in 1992, 350 years after Galileo's discoveries, finally vindicated him in a statement apologizing for the Catholic Churches' treatment towards Galileo.

35. The first person to determine the law of gravity and explain the motion of planets was Sir Isaac Newton.

36. Apart from visible light that we see, the Milky Way gives out many types of energy, including infrared light, gamma rays, dark matter, radio waves, and x-rays are emitted from our galaxy.

37. Every 200 years, a lunar eclipse will happen three times in the same year.

38. The distance between Earth and our moon would fit all the planets lined up together.

39. Mike Brown's team, who discovered Haumea, first

saw it just after Christmas on 28 December 2004 and called it 'Santa.'

40. The strongest Martian winds happen when Mars is closest to the Sun.

41. A massive group of stars, star clusters, interstellar gas and dust, and dark matter which is held together by gravity is called a galaxy.

42. The International Space Station (ISS) weighs almost 419,500 kg (925,000 lb).

43. The International Space Station (ISS) is a sizeable human-made spacecraft that orbits Earth. Humans live and research the spacecraft in space.

44. Mars' moons are named after the twin gods for Panic and Fear, who went into battle with Mars.

45. The dwarf planets, Eris and Pluto, are both smaller than Earth's moon.

46. Pluto's four small satellites that orbit it are called Nix, Hydra, Kerberos, and Styx.

47. Comets and asteroids are alike, but asteroids don't have a fuzzy outline and tail as comets do.

48. Pluto is the largest dwarf planet.

49. While most planets don't have seasons, Uranus does.

50. A constellation is a group of stars in the sky that has been grouped to make a pattern. Patterns include animals, mythological creatures, and objects.

51. As of January 2019, over 61,000 meteorites have been found on Earth, of which 224 are from Mars.

52. The Magellanic Clouds belong to the Local Group and orbits the Milky Way.

53. There is an ISS water recovery system that reduces water delivery dependence by 65%.

54. Charon is half Pluto's size, about half the width

of the United States, and just bigger than India. Because of this, Pluto and Charon are often referred to as the double dwarf planet system.

55. Scientists believe that in the Milky Way, seven new stars are formed every year.

56. Software on the ISS, including 350,000 sensors, was designed to provide health and safety monitoring for the crew and station.

57. Jupiter's rings are thick, colorful clouds of deadly poisonous gases.

58. One of the prerequisites for extraterrestrial life is liquid water.

59. Modern telescopes can detect infrared and radio waves.

60. The chemical element Plutonium was named after Pluto in 1941.

61. Infrared astronomy is the detection and analysis of infrared radiation.

62. The ancient Babylonians and Far East astronomers first observed Saturn.

63. A flyby mission is estimated to take about 16 years to Makemake with assistance from Jupiter's gravity.

64. A lunar eclipse lasts longer than a solar eclipse.

65. A small star will live longer, up to hundreds of billions of years, than a giant star which will only live for a few million years. Our sun is medium-sized and will shine another 5 billion years.

66. Millions of meteors enter the Earth's atmosphere every day.

67. The first reusable space equipment was the space shuttle.

68. Gravitational interactions with two companion galaxies called M32 and M110 are distorting the spiral arms of Andromeda Galaxy.

69. The first person to observe Mars through a telescope was Galileo Galilee in 1609.
70. In 2006, Ceres, Pluto, Eris, Makemake, and Haumea were all given the status dwarf planet.
71. The side of the moon that is visible to Earth at any particular time is called the Near side of the moon. The other far side is sometimes called the Dark side even though it isn't as it faces the sun.
72. The first planet to be discovered through a telescope is Uranus.
73. Radio waves, which can travel through space, are used by astronauts that are in space to communicate to Earth.
74. The moons of Jupiter are sometimes called the Jovian satellites, the biggest of which are Ganymede, Callisto Io, and Europa. Ganymede measures 5,268 km across, making it bigger than Mercury.
75. In comparison to the Messier 87 with a diameter of 980,000 light-years and Hercules A with a diameter of 1.5 million light-years, the Milky Way is very small (100,000 light-years).
76. The moon is 81 times lighter than Earth.
77. Mars has been around for 4.5 billion years.
78. Streams of hydrogen gas and embedded stars connect the Triangulum Galaxy and Andromeda galaxies. It is suggested that these two galaxies will interact again in about 2.5 billion years.
79. In 1976, Viking I produced a photograph of a rock on Mars with a human face. Many people believed that extraterrestrials created the face. The spacecraft 'Global Surveyor' confirmed that the face was an optical illusion. However, the people who believed in the extraterrestrials accused NASA of changing the data.

80. You can see the Whirlpool Galaxy and its companion NGC 5195 with binoculars.

81. Mercury is one of the five planets you can see without a telescope. The others are Mars, Jupiter, Saturn, and Venus.

82. Venus rotates slower than Earth on its axis, so a day on Venus would be equivalent to about 243 Earth days.

83. Haumea has two moons, Namaka and Hi'iaka, who were named after the goddess of childbirth.

84. The distance from Earth to the Sun is approximately 149,597,891 km (92,955,820 mi).

85. We can see Halley's Comet every 75 to 76 years as it nears Earth. The last time we saw Halley's Comet was in 1986, and the next time is expected to be in the year 2061.

86. Earth's rotation slows down about 1.5 milliseconds every 100 years due to the moon's gravitational force. The same effect puts the moon into a higher orbit by about 3.8 cm or 1.5 in every year.

87. Of the five dwarf planets, only two have been visited by space probes. In 2015 NASA's Dawn and New Horizons reached Ceres and Pluto.

88. Mercury's surface is like the surface of our moon, with a barren, rocky surface, and lots of craters.

89. The Big Dipper and the Little Dipper are not constellations. They are patterns within constellations (known as asterisms). For example, the Big Dipper is an asterism contained in the Ursa Major constellation.

90. The distance from Saturn to the sun is 1,425,725,413 km (885,904,700 mi).

91. Some believe that one of Jupiter's moons, Europa, may have life.

92. Throughout history, stars have had very important

roles. They were used to assist navigation, became part of many religious practices, and used in astrology.

93. Mercury is the fastest planet that orbits the Sun, taking 88 Earth days to complete one orbit.

94. The Space Shuttle was NASA's transportation system in space. It transported astronauts and cargo to and from Earth's orbit.

95. One of Saturn's moons, Pan, absorbs some of the material making up Saturn's rings, making it walnut-shaped.

96. Elliptical galaxies contain not much gas or dust, so very few stars are formed. They are often the largest galaxies and the oldest.

97. Liquids form into spheres in space instead of flowing like on Earth.

98. The Pinwheel Galaxy is a beautiful spiral galaxy with a stunning pinwheel structure.

99. Almost anyone who travels into space can be called an astronaut or a cosmonaut.

100. Meteorites that have crashed on Earth have contained small traces of Martian material in them. This has enabled our scientists to study Mars in greater detail.

101. As of February 2020, the ESA's Space Debris Office has identified 34 000 objects >10 cm and 900 000 objects from 1 cm to 10 cm occupying space. There are over 128 million objects of space debris with a size of 1 mm to 1 cm!

102. A lunar eclipse can be partial, full, or penumbral.

103. Perihelion is the closest point in a comet's orbit to the Sun, while aphelion is the furthest point.

104. The dwarf planet Ceres has similar surface features, as well as a rocky core, to inner planets.

105. Several craters in Charon were named after Star

Wars and Star Trek characters, Darth Vader, James T Kirk, Spock, and Uhura.

106. Venus has a perfectly circular sphere.

107. Our solar system lies about 27,000 light-years from the Galactic Centre of the galaxy, within the disk of the Milky Way Galaxy.

108. Mike Brown, the person who first discovered Eris, the dwarf planet, wanted to name it Lilah after his new baby. In the end, he didn't call it this, as it would have been controversial for the rest of his family.

109. In 1610, Galileo Galilei discovered the four largest moons of Jupiter with a telescope.

110. The mass of Neptune is 17.15 times Earth's mass, at 102,410,000,000,000,000 billion kg.

111. Even though Uranus is four times bigger than Earth, you need a telescope to see it.

112. Eris and its moon Dysnomia are the furthest known objects in our solar system.

113. When a moon is first discovered, it is given a provisional designation until the IAU approves an official name and confirms the discovery of the moon.

114. The ancient Greeks used to think Mars revolved around Earth as they believed that Earth was the center of the universe.

115. The M87 Galaxy was named after the person who discovered it, Charles Messier. The number refers to the 87th member in his catalog.

116. Experimental and spacecraft systems on the ISS make up an area of around 100 telephone booths.

117. The inner planets don't have any rings around them.

118. 99% of the human body is made up of hydrogen, nitrogen, carbon, and oxygen. The hydrogen

atoms came from the Big Bang, and the nitrogen, carbon, and oxygen were made from burning stars.

119. Uranus is named after the Greek god Ouranos. All other planets are named after Roman gods.

120. There is no center of the Universe!

121. The biggest artificial satellite that orbits Earth is the International Space Station.

122. The Oort Cloud is shaped like a sphere and has an outer spherical shape and an inner disk shape.

123. The name Eris was accepted by the IAU on 13 September 2006.

124. Ganymede, Titan, Io, Callisto, Triton, Europa, Mercury, and our moon are all larger than Pluto.

125. Scientists think that Venus rotates backward from an asteroid collision in the past.

126. Inner planets, i.e., planets closest to the sun, consist mostly of an iron core surrounded by a mantle.

127. Callisto, one of Jupiter's moons, is the same size as Mercury, but it is not a planet as it orbits Jupiter and not the Sun.

128. Yuri Gagarin, a Russian, was the first person in space. In 1961 he orbited Earth in 108 minutes on the spaceship Vostok 1.

129. Rockets launch artificial satellites into orbit in space.

130. The Solar system is situated on the Orion Spur, which is an arm of the major Sagittarius Arm of the Milky Way.

131. The Milky Way is called the Milky Road by the ancient Romans as it reminded them of milk.

132. Halley's Comet has been observed for thousands of years. Babylonian, Chinese, and European

stargazers have all recorded observations of Halley's Comet.

133. Buzz Aldrin hit a man in the face in 2002 when the man claimed the moon landings were fake.

134. The fifth-largest moon in the solar system belongs to Earth.

135. Reflecting telescopes uses concave mirrors where one mirror captures an image, and a second mirror reflects the image from the first image. This is also called a Newtonian telescope.

136. Earth has a powerful magnetic field caused by its nickel-iron core together with its fast rotation.

137. It takes the same time for the moon to rotate on its axis as it does for it to orbit Earth. This means we can only see about 60% of the moon's surface from Earth.

138. There are six different types of observational astronomy: gamma-ray astronomy, optical astronomy, radio astronomy, ultraviolet astronomy, infrared astronomy, and x-ray astronomy.

139. More than 230 people from 18 countries have already visited the ISS.

140. The Kennedy Space Centre in Florida was the launch pad of all the space shuttles.

141. When two pieces of the same metal contact each other in space, they join together and are "cold-welded."

142. Three men from Yemen claimed they inherited Mars 3,000 years ago, so in 1997, they sued NASA for trespassing on Mars!

143. The number of water molecules in 10 drops of water is equivalent to all the stars in the Universe.

144. On 19 January 2006, NASA launched the New Horizons spacecraft. It flew within 27,359 km

(17,000 mi) of Charon, Pluto's largest moon, nine years later, on 14 July 2015.

145. High altitude balloons were first used to test outer space exploration.

146. Charon, one of Pluto's moons, appears to have ice-based geology with active ice geysers and ice volcanoes.

147. The Spitzer Space Telescope is orbiting at around 26 million miles from Earth.

148. In Greek mythology, Hercules had to fulfill 12 tasks or labors for killing his family. One of his jobs including killing a giant crab, which he then kicked to the stars, thus forming the Cancer constellation. Other versions of the story are that he crushed the crab with his foot or he killed it with his club.

149. The recording of data through observation with tools such as telescopes is called observational astronomy.

150. Asteroids are made up of precious and non-precious metals and water.

151. Neil Armstrong, from the Apollo 11 mission, was the first man to step on the moon.

152. One Mercury year equals 88 days on Earth.

153. The Big Bang theory is based on the inflation theory, i.e., the universe suddenly expanded, doubling in size every 10-34 seconds. The 'bang' explosion lasted about 10 - 30 seconds but it changed the hand-sized universe to one that is now 10,000,000,000,000,000,000,000,000 times bigger.

154. When the core of a star collapses after a supernova explosion, neutron stars are created. They may be small (about 10 km (6 mi) in radius,

but they can spin very fast at about 600 - 712 times per second.

155. An American poll taken in 2019 showed that 6% of people thought the moon landing was fake while 15% sat on the fence and didn't know if it was false or if it happened.

156. To date, NASA knows no astronaut that has had sex in space! Getting pregnant would also be very dangerous in space.

157. Callisto, Jupiter's second-largest moon, has the most craters and dates back to 4 billion years ago when the Solar System was first formed. It is one of the oldest landscapes in our Solar System.

158. The diameter of the Whirlpool Galaxy is about 76,000 light-years and lies about 23 million light-years from Earth.

159. The Pinwheel Galaxy's star birth regions contain lots of hydrogen, so they are called HII regions.

160. Jose Luis Ortiz Moreno from Spain claimed he discovered Haumea in 2005, but Mike Brown and his US team made the same claim in 2004. In 2008 the IAU recognized Mike Brown's discovery after determining the Spanish club may have committed fraud.

161. The International Space Station (ISS) is located in the ionosphere, the first part of outer space.

162. Pluto is located on average 5,906,380,000 km (3,670,050,000 mi) from the Sun.

163. A dwarf planet orbits the Sun and has gravity and is not a satellite or a moon.

164. Dark Matter is the matter that holds stars and dust together in a spiral galaxy. It makes up most of the mass.

165. In 1959, the first US monkeys who survived a trip

to space, on a Jupiter rocket, were Able and Miss Baker.

166. Luna 1, the first space probe that was sent to the moon, missed it by about 5,000 km (3,000 mi).

167. Jupiter has a strong magnetic field! You would weigh 2.5 times more than you would on Earth.

168. Many scientists believe that Saturn's moon Titan has similar conditions to Earth's early years, except temperature due to its distance from the Sun.

169. In 2008 Haumea was classified as one of the dwarf planets.

170. The Hubble Space Telescope can observe many things but not the Sun or Mercury.

171. Mars came closest to Earth on 27 August 2003. The next time it will happen, this close will be in the year 2287.

172. One of Mars' moons, Phobos, is slowly merging with the Red Planet and will crash in about 50 million years, creating rings around Mars.

173. Six spacecraft may be connected to the ISS at one time.

174. Ceres is situated between Mars and Jupiter in the asteroid belt.

175. Makemake, dwarf planet, was discovered just before Easter and so was named after the god of fertility from Rapa Nui, natives of Easter Island. Before publicizing the discovery, the project team used a secret codename "Easter Bunny."

176. Venus has no moons or rings.

177. Asteroids, made of rock and metal, are small Solar Systems that circle the Sun. Some scientists think that the organic compounds found in asteroids created life on Earth.

178. A modern spacecraft would take approximately

450 million years to fly to the center of the Milky Way.

179. Some scientists believe that in the future, the Sun will move and join the Andromeda Galaxy.

180. Mars appears to move backward through the zodiac, so the ancient Egyptians used to call it the "backward traveler."

181. When you look at the Milky Way at night, you only see about 0.0000025% of the billions of stars in the galaxy.

182. There are three types of meteorites - stony, iron, and stony-iron.

183. The Milky Way rotates at a very fast speed of 270 km per sec (168 miles per sec). So where you were an hour ago is 965,606 km (600,000 mi) away.

184. 5 European explorers and astronomers named 40 out of the 88 constellations (48 were appointed by Ptolemy). Their names are Gerardus Mercator (Dutch explorer in 1500s), Pieter Keyser and Frederick de Hautmann (beginning of the 16th century). Johannes Hevelius (1690), and Nicolas Louis de Lacaille (French astronomer in 1750s).

185. With technological advances, astronomers now have powerful telescopes for their research.

186. The most famous Comet is named Halley's Comet.

187. Pioneer 10 was the first spacecraft to get up close to Ganymede on a mission to Jupiter. In 1979 Voyager 1 and Voyager 2 found that Ganymede was larger than Titan, Saturn's moon, which at the time was considered to be more significant. In 1996 the Galileo spacecraft flew by Ganymede and discovered its magnetic field.

188. Valles Marineris meaning Mariner Valley is a huge Martian canyon measuring 4,023 km (2,500

mi) long and 6.4 km (4 mi) deep. It is the longest valley in the solar system and stretches a distance similar to New York to San Francisco.

189. The fuzzy tail of a comet is called a coma. This happens when the Comet nears the sun, and the sun's heat melts the frozen water from the surface of the Comet.

190. Titan (one of Saturn's moons) is made up of a rocky core that is surrounded by layers of water ice. Even though ice is surrounding it, the core is still boiling and full of liquid water and ammonia.

191. The Triangulum Galaxy is a spiral galaxy without a central bar. It has loosely wound arms that are attached to its galactic core.

192. The highest mountain on Venus is the Maxwell Montes, about 8.8 km (5.4 mi) high, similar to Mt Everest, the highest mountain on Earth.

193. Europa, Jupiter's Earth, is covered in ice that is about 100 km (62 mi) thick.

194. The sun has burnt off half its hydrogen store and will continue to burn for another 5 billion years when it will die.

195. Some meteoroids travel at speeds of 42 km/sec (26 mi/sec) through our solar system.

196. The Oort Cloud is a theoretical cloud beyond the edge of our solar system.

197. The airflow around an astronaut's head is maintained to avoid a bubble of carbon dioxide forming around their head.

198. After Saturn, Uranus is the next least dense planet.

199. You can sit inside some very large reflecting telescopes.

200. The Royal Family paid William Herschel (the

person who discovered Uranus) 200 pounds to look through his telescopes.

201. About once weekly, the sun rises and sets on Pluto.

202. The Sun's inner core is about the same temperature as Earth's inner core.

203. Scientists believe that volcanic processes created Venus.

204. A nebula that has no well-defined boundaries is called a diffuse nebula. The Carina Nebula is an example of a diffuse nebula.

205. There are online sites that you can find that lets you view the night sky through the lens of a live telescope.

206. In ancient times during a solar eclipse, the Chinese believed that an enormous dragon swallowed the sun, and they made as much noise as possible to try and scare the dragon away.

207. A group of stars in the middle of an elliptical galaxy can often look like one bright star.

208. Using the Hubble Space Telescope, astronomers discovered Xanadu on Titan in 1994, which is a highly reflective area about the size of Australia. Radar images depict dunes, hills, valleys, and rivers!

209. Siding Spring, an Oort Cloud comet that was observed close to Mars in January 2014, won't return to our Solar System for another 740,000 years.

210. The first spacecraft to successfully land on Mars, on 20 July 1976, is NASA's Viking 1.

211. The pictures from a refracting telescope are blurrier than those from a reflecting telescope.

212. Elliptical galaxies are typical in galaxy clusters.

213. From 1989 to 1994, one of the largest storms ever,

known as the Great Dark Spot, was observed on Neptune.

214. A famous TV show host, Bill Nye, has applied and rejected many times to be an astronaut with NASA.

215. An international research team led by NASA in November 2019 detected water vapor for the first time above Europa's surface using a spectrograph at the Keck Observatory in Hawaii. If scientists could study the composition of these plumes, it will help them to determine if life is possible on Europa.

216. A hurricane-like vortex covers Saturn's south pole.

217. Full moons vary in size, depending on whether it's far away or closer to Earth.

218. A solar eclipse occurs when the moon completely covers the sun.

219. Buzz Aldrin once quoted 'Mars is there, waiting to be reached.'

220. Ceres, with a diameter of 950 km, is now a dwarf planet. However, it was first known as an asteroid in 1801 when Giuseppe Piazzi found it.

221. Neptune has 13 moons, of which Triton is the largest.

222. Due to its proximity to the Sun, the Sun's rays are seven times stronger on Mercury than on Earth.

223. Deimos is one of Mars' moons. It means to dread or terror. The other moon is named Phobos, which means fear.

224. Theoretical astronomy involves using analytical models to study topics such as stellar dynamics, galaxy formation, matter in the universe, the origins of cosmic rays, evolution, general relativity, and astroparticle physics.

225. When an asteroid collides with a planet, the result

is called an impact event. An example of this is the extinction event 65 million years ago that made dinosaurs extinct.

226. The eight planets in order of their distance from the sun are Mercury, Venus, Earth, Mars, Jupiter, Saturn, Uranus, and Neptune.

227. Explorer 1, USA's first artificial satellite, was launched three months after Sputnik 1.

228. Total solar eclipses cannot be observed at the north or the south poles.

229. About every 2,000 years, a space object about the size of a football field hits Earth.

230. Sally Ride, the first American woman to go to space in 1983, was also the youngest American to go to space.

231. Most of the universe is made up of dark matter, according to scientists.

232. The Martian movie paid homage to the Greek God of war, Ares (known as Mars), when they named the mission Ares 3.

233. Scientists believe that Andromeda was created when small galaxies collided about five to nine million years ago.

234. Objects sucked into black holes are torn apart due to the strong gravitational pull.

235. As a result of the Earth's gravitational pull, moonquakes occur (like Earthquakes but on the moon).

236. Astronomy, which examines the position of space objects, differs from astrology, which is a belief system for human behavior.

237. Pluto's orbit around the sun is elliptical, traveling at a speed of about 16,809 kph (10,444 mph).

238. The space station completes 15.5 orbits around Earth in a day, which means that every 92

minutes, crew members onboard the station experience a sunrise or sunset.

239. Most telescopes can detect some sort of electromagnetic waves.

240. Venus has few small crater impacts than other inner planets as its thick atmosphere, 92 times greater than Earth's, protects it from meteors and asteroids that enter its atmosphere.

241. The largest impact crater on Mercury, the Caloris Basin, has a diameter of 1,550 km (963 mi). The Mariner 10 probe found it in 1974.

242. The densest planet in the Solar System is Earth.

243. The size of meteoroids can be as small as dust and as big as 10 meters (32.8 ft) in diameter.

244. Galileo is often thought of as the inventor of the telescope. However, this is incorrect! Hans Lippershey invented the telescope in 1608. Galileo was, however, the first person to use it to study the night sky.

245. Due to its rapid rotation, Haumea is one of the densest dwarf planets.

246. The Hubble Space Telescope completes 15 orbits around Earth every day—approximately one every 95 minutes.

247. Only telescopes that detect radio waves, infrared light, and x-rays can be used to study the Milky Way as optical telescopes can't see through the dense gas and dust.

248. Due to the amount of hydrogen sulfide in its atmosphere, Uranus may have a rotten egg smell.

249. Astronomers can't understand why the universe bends, so if you walked in a straight line out into the universe, you would come back to where you started.

250. Astronomers have found two bright spots on a

crater on Ceres at about 19 degrees latitude, which they do not think came from a volcano. Researchers believe it may be ice or salts.

251. Hydra is the largest constellation and covers 3.1% of the night sky.

252. The Oort Cloud is located beyond the Kuiper Belt, past Pluto. It is a big thick bubble around our solar system filled with comets.

253. A reflection nebula can't emit its light and therefore reflects the colors of other nearby stars. It often looks blue. Usually, an emission nebula is nearby if a reflection nebula exists.

254. On March 18, 1965, Alexei Leonov from Russia was the first person to perform a spacewalk on the Voskhod 2 mission.

255. About 400 active satellites orbit above Earth at a distance greater than 35,786 km (22,236 mi) above the Earth's surface. These are known as High Earth Orbit satellites.

256. A black hole is about the size of a massive star.

257. More than 50,000 meteorites have been found on Earth, of which 99.8% of them come from asteroids.

258. Venus spins in the opposite direction to other planets. Uranus also rotates in the opposite direction.

259. In 2004, another dwarf planet now called Sedna was discovered beyond Neptune. Sedna orbits the sun in one of the coldest known regions of our solar system and so was aptly named after a goddess who lived at the bottom of the frigid Arctic ocean. Sedna takes 10,500 years to complete its orbit around the sun and never enters the Kuiper Belt. Astronomers have suggested that

Sedna is the first observed object that belongs to the inner Oort Cloud.

260. Pressure systems in the atmosphere create our weather. Air movements are caused by air temperature and pressure changes, which in turn creates wind.

261. Four times Earth's size, Neptune is a large planet.

262. The month of March is called after the planet Mars.

263. While there is no evidence of life on Venus, scientists doubt any life could exist due to its high temperature.

264. The Whirlpool Galaxy is also called Messier 51a, M51a, and NGC 5194. It has a companion galaxy, NGC 5195 that is sometimes just called M51.

265. During half of its orbit, Uranus receives direct sunlight for about 42 years and dark for the next 42 years.

266. Jupiter is the largest planet in our Solar system and is so big that it could fit 1,300 Earths inside it.

267. In 1668, Sir Isaac Newton invented the reflecting telescope, but it only became a popular tool in Astronomy around 100 years later.

268. Jupiter orbits the Sun once every 11.8 Earth years.

269. A gas giant is a large planet made mostly of hydrogen and helium. Jupiter and Saturn are two gas giants in our solar system. Uranus and Neptune are also considered to be gas giants as they are made of heavy unstable substances.

270. Over 200,000 people have applied to fly to Mars and never return! Of these, 100 were selected, and 24 will make the trip.

271. Once, an astronaut was almost turned down to become an astronaut as he suffered from hay

fever. NASA then realized that there's no pollen in space!

272. Some scientists believe that we could only see the universe under a microscope about 10 - 43 seconds after Big Bang. The number 10 - 43 is equal to 0.0000000000000000000000000000000000000000001, or one 10 million trillion trillionths of a second.

273. The first African American woman in space was Mae Jemison onboard the Endeavour on 12 September 1992.

274. Planets that can be seen at the time of a total solar eclipse will be seen as points of light in the sky.

275. Haumea orbits the sun in 285 Earth years.

276. S Andromedae, a supernova in the Andromeda Galaxy, was recorded in 1885 by spotting it through a telescope.

277. Chinese and Islamic astronomers recorded a supernova that could be seen in the daytime in 1054.

278. A sonic boom is usually heard seconds after you see the meteor, but only if the meteor is making a sound.

279. Gas giants used to refer to all giant planets.

280. The Large Magellanic Cloud is located about 163,000 light-years from Earth, while the Small Magellanic Cloud is further away at about 200,000 light-years.

281. Lippershey was the first person to apply for a patent on a telescope in 1608. In 1609, Galileo made his telescope and was the first to look through it into space. He could see mountains and craters on the moon, as well as the Milky Way. He also discovered the rings of Saturn, sunspots, and four of Jupiter's moons.

282. About one septillion (1, 000, 000, 000, 000, 000, 000, 000, 000 or a trillion) snow crystals drop from the sky every winter.

283. Pluto was closer to the sun than Neptune for 20 years of its 248-year orbit. It will be closer to the sun again in 2231.

284. Crux, also known as the Southern Cross, is the brightest constellation.

285. An artificial satellite is made by humans.

286. The dwarf planet Eris was named after the Greek goddess of strife and discord.

287. Deimos' (one of Mars moons) temperature is about -40.15 degrees C (-40.27 degrees F).

288. Some scientists believe that over 300 million years ago, Venus may have had oceans on its surface, but they dried up when the Sun's rays became stronger.

289. The Oort Cloud is named after a Dutch astronomer Jan Oort.

290. Deimos, one of Mars moons, has a radius of 6.27 km (3.9 mi). It is very small, with many craters, and it takes about 30.3 hours to orbit Mars. Deimos shines as bright as Venus in the full moon and becomes one of the brightest objects in the sky.

291. Mars' environment resembles the Antarctic deserts on Earth.

292. Mars' winds can blow at up to 201 kph (125 mph) and can cover the whole planet.

293. The Hubble Space Telescope gets its energy from the Sun through solar panels.

294. On 20 August 1909, The Yerkes Observatory took the first pictures of Pluto. However, at the time, astronomers didn't know it was Pluto.

295. Charles Messier discovered the Triangulum

Galaxy in 1764 and named it as object 33 or M33.

296. No plans have been announced from India or Japan for a human trip to Mars.

297. An emission nebula that is formed by glowing and expanding ionized gases is called a planetary nebula. The Cat's Eye Nebula is an example of a planetary nebula.

298. The compound uranium was named after the planet Uranus.

299. Only one probe has successfully landed on Titan - many have been very close.

300. The Hubble Space Telescope observed a moon orbiting Makemake in 2015. This moon has a provisional name, S/2015 (136472) 1, and is nicknamed MK 2.

301. Scientists believe that Mars is probably the most 'liveable' planet in our solar system.

302. When the moon is in the New Moon phase, it doesn't show in the sky (no moon) as the sun is shining on the other side.

303. The largest crater on Deimos, one of Mars' moons, is Voltaire, with a diameter of 3,057 km (1,900 mi). Voltaire was named after Francois-Marie Arouet, a French writer who had a pen name, Voltaire.

304. The sixth planet from the sun is Saturn.

305. No one knows who discovered Mercury, but it was first observed through a telescope in 1631 by Thomas Harriott and Galileo Galilee.

306. Nearly four billion years ago, asteroids crashed into Mars, creating a large plain the size of the Caribbean Sea (2,092 km or 1,300 mi). Hellas is the name of this plain.

307. Many astronomers believe that black holes exist in

the center of all galaxies, including the
Milky Way.

308. One Uranus day is equivalent to 17 Earth hours
and 54 minutes.

309. Scientists give many nicknames to rocks on Mars,
so they can easily remember them. Some of these
names are Barnacle Bill, Shark, Moe, Pop Tart,
and Cabbage Patch.

310. Pluto is 456 times smaller than Earth.

311. No space shuttles occurred between December
and January as the computer couldn't handle a
change in the year!

312. A nebula is a giant cloud of dust and gas in space
formed by the explosion of a dying star, for
example, a supernova, or where new stars are
being created.

313. Asteroids are also known as minor planets or
planetoids.

314. Scientists estimated that the sun would become a
nebula in about 5 billion years.

315. Venus has no liquid water.

316. During Jupiter's and Saturn's lightning storms,
methane is converted into carbon soot, which then
hardens into graphite and then diamonds. This
means that it rains diamonds, sometimes up to a
centimeter (0.39 in) thick, on Saturn and Jupiter.

317. The four moons of Pluto are Nix (the Greek
goddess of night and darkness), Charon (ferryman
of Hades), Hydra (the nine-headed serpent who
guards Hades) and S/2011 P 1 (named after the
year it was found).

318. A galaxy that is very small with about 10 million
stars is called a dwarf galaxy.

319. Lyman Spitzer was the first person to suggest
launching a telescope into space.

320. Ultraviolet astronomy is the observation of ultraviolet wavelengths from space or the Earth's upper atmosphere.

321. The diameter of Mercury is 2/5 that of Earth's.

322. Space is completely silent as there is no atmosphere for sound to travel through.

323. Every day, hundreds of tons of material from asteroids and comets fall toward Earth, but they do not reach Earth. They are destroyed as they pass through our atmosphere. They are then called a meteorite when reaching Earth.

324. Mars has a similar size and landmass to Earth.

325. Elliptical galaxies are listed as type E, followed by a number that represents the degree of their ellipticity.

326. A full lunar eclipse happens when Earth passes directly in front of the Moon. Halfway through the eclipse, the Moon may appear blood red in color.

327. When an astronaut is selected to be a NASA trainee, they train for 20 months. They often train underwater in swimming pools to test equipment and simulate spacewalks.

328. Giant galaxies in the universe can be as long as 2,000,000 light-years.

329. President Woodrow Wilson started the National Advisory Committee for Aeronautics (NACA) before NASA came about in 1958.

330. It took Apollo 11 four days and six hours to reach the moon.

331. When the gases and dust from a nebula become very big, it starts to collapse due to their strong gravity. This causes the material to heat up at the center of the nebula. The hot center is the beginning of a new star.

332. Stephen Hawking believed that increasing temperatures cause black holes to shrink and evaporate.

333. The Milky Way was the only galaxy we knew until 1924 when Hubble proved there were other galaxies in the universe.

334. In Chinese mythology, the gods placed the Milky Way (which they called the Silver River) in the heavens to separate a weaver from a herdsman, a couple in love.

335. A comet is known as a short term comet if it has an orbital period of fewer than 200 years. A long term comet has an orbital period of more than 200 years.

336. Mars' sunrises and sunsets look blue as there is almost no atmosphere on Mars.

337. Scientists believe Haumea and the sun were created about the same time, 4.5 billion years ago.

338. No one has ever stepped on the far side of the moon.

339. About 8,000 light-years away, you can see the Hourglass Nebula, which is called this due to it resembling the shape of an hourglass.

340. The rate of skin cancers increases because the ozone layer in the atmosphere is decreasing and unable to absorb the ultraviolet rays from the sun.

341. The Milky Way will collide with the Andromeda galaxy in about 4.5 billion years.

342. The Milky Way has already consumed smaller galaxies. It continues to pull matter away from the Large and Small Magellanic Clouds.

343. The word "Friday" means "Venus day."

344. Our Sun is believed to be an average-sized star, but it is so big that you could fit 1,000,000 planet Earth inside it.

345. Some asteroids have moons.

346. The gravitational pull of the moon has slowed Earth's rotation. Our days used to be shorter.

347. Mariner 9 was the first spacecraft to orbit Mars in 1971. Although many photographs were taken of Deimos, it didn't land on the moon.

348. NASA's Hubble Space Telescope was launched on April 24, 1990, from the Kennedy Space Centre by the Space Shuttle.

349. Jupiter has a system of thin rings. Its rings are mainly dust particles from impacts of comets and asteroids coming from some of Jupiter's smaller worlds.

350. At one time, the oldest known map of the moon was drawn by Leonardo da Vinci about the year 1505 until an older one was found, carved into a prehistoric tomb in Ireland that was about 5,000 years old!

351. Neptune has no solid body (it's gas), so its equatorial clouds rotate around it in 16 hours.

352. The second closest star to Earth, after the sun, is Proxima Centauri.

353. Venus' diameter is about 12,104 km (7,521 mi), which is nearly the same size as Earth's.

354. The age of the Milky Way is about 14 billion years.

355. Charon, Pluto's moon, can be pronounced "SHAR-on" or "CARE-on." Both pronunciations are correct.

356. Lb', the abbreviation for pound, means 'balance' from Libra, an astrology sign.

357. Charon has a northern region named Mordor after a region in the Lord of the Rings. It is darker than the rest of the moon from small particles of tar from Pluto's atmosphere.

358. Ganymede, one of Jupiter's moons, is named after a mythical Greek boy who, disguised as an eagle, was carried to Olympus by Zeus. He became the Olympian gods' cupbearer.

359. Makemake travels through the Kuiper Belt like all the dwarf planets except for Ceres.

360. One year after the Soviet satellite Sputnik 1 was launched, NASA was established.

361. A full moon appears bright but reflects only about 7% of the sun's rays.

362. The Pinwheel Galaxy has an unusual pattern resembling a pinwheel that scientists believe energetic dust and gas winds created.

363. Jupiter's surface temperature is -108 degrees celsius (-162.4 deg F).

364. Homer wrote about Orion, Bootes, and Ursa Major constellations in his poem Illiad and the Odyssey in the 8th century BC.

365. On October 31, 2000, the first ISS crew mission called 'Expedition 1' was launched on a Russian Soyuz. On November 2, 2000, the 3 Russian cosmonauts docked in and entered the ISS. Since then, the space station has been occupied continuously, making it the most extended continuous human presence in space.

366. Stephen Hawking once said, "Earth might one day soon resemble the planet, Venus."

367. A spiral galaxy is the most common type of galaxy.

368. Antarctica was discovered before Uranus.

369. As of 2013, 76 people visited the ISS on two occasions; 25 people made three trips to the station, and five people have been four times.

370. When a galaxy merges with another, a galactic merger is formed.

371. Some of the most densely populated globulars are amongst the 450 globular clusters that orbit around the Andromeda Galaxy.

372. In the Cancer Constellation, the star Asellus Australis is called 'southern donkey.' In contrast, the star Asellus Borealis is called the 'northern donkey.'

373. All planets are named after a God except Earth.

374. The Spitzer Space Telescope was a very successful reflective telescope and captured light images of planets outside of our Solar System.

375. There's a rocky core at the center of Jupiter, slightly bigger than Earth, but weighing about 20 times as much.

376. Neptune is called after the Roman god of the sea.

377. Neptune has 14 moons in orbit; the largest one is as large as Pluto!

378. As of 2013, Anatoly Solovyev holds the record from having made the most spacewalks - a total of 16 spacewalks over 82 hours and 22 minutes.

379. The far side of the moon is thicker than the near side.

380. The International Space Station (ISS) is the largest artificial body in orbit. It is 109 m (357 ft) in length, which makes the area of the space station span about the size of an American football field.

381. Most of the modern telescopes are reflectors.

382. A 50 kg (110 lb) person on Earth would weigh 1.5 kg (3.3 lb) on Charon. A 50 kg (110 lb) on Charon would weigh 1,751 kg (3,853 lb) on Earth.

383. Callisto has been considered to be the most suitable place for a human base of further exploration as it is not geologically active, and life may be possible in its underwater ocean.

384. The Sombrero Galaxy is also called Messier 104, M104 or NGC 4594.

385. Supernovas are one of the most violent naturally occurring phenomenons in outer space.

386. Charon is 19,640 km (12,203 mi) distance from Pluto, which is closer than our moon is to Earth, 386,000 km (239,000 mi) distance.

387. A black hole 10 billion light-years away contain a giant cloud of water vapor, 140 trillion times the amount of water that Earth has in its oceans. This is the biggest discovery of water found in space.

388. The stratosphere is the second closest atmospheric layer to Earth and is about 50 km (31 mi) from the Earth's surface.

389. Makemake, dwarf planet, has no atmosphere.

390. There were many flyby missions to Jupiter. One orbited the planet successfully from 1995 to 2002.

391. Between 3 to 4 million years ago, meteorites bombarded the moon, a phenomenon known as 'lunar cataclysm.'

392. It is thought that dinosaurs became extinct when a 12.8 km (8 mi) long meteor hit Earth.

393. Eris, dwarf planet, rotates on its axis every 25 hours (similar to Earth's 24 hours). It orbits the sun every 557 years.

394. About 80% of the weight of Earth's atmosphere is located in the closest and thinnest layer, the troposphere.

395. The Local Group is a group of galaxies, including the three largest, Andromeda, the Milky Way, and the Triangulum Galaxy. It is about 10 million light-years in diameter. In addition to the three mentioned, it contains about another 30 smaller galaxies.

396. Makemake is a dwarf planet and the third-largest after Pluto and Eris.

397. There are no large mountains or volcanoes on Callisto, Jupiter's second-largest moon, only impact craters and multi-ring structures. The size of the impact craters can range from 0.1 km (0.06 miles) to over 100 km (60 miles).

398. A white hole looks like a black hole, spinning with rings of dust and gas around the event horizon. They differ, however, to black holes as they spit something out instead of absorbing everything that goes into it. Nothing can go into a white hole.

399. When Phobos, Mars' largest moon, is destroyed, it will give Mars rings like other planets.

400. An astronaut has to use about 70 to 110 tools to complete tasks on their spacecraft when they perform an extra-vehicular activity or moonwalk.

401. Pluto has the same age as the solar system, i.e., 4.6 billion years.

402. An astronaut that reaches more than 100 km (62 miles) in altitude is recognized by the World Air Sports Federation (FAI). Astronauts who reach 80 km (50 miles) in the USA are awarded astronaut wings.

403. Dysnomia orbits its dwarf planet Eris every 16 days.

404. The gas that flows out of the Magellanic Clouds is being absorbed by the Milky Way and may collide or merge with it in the end.

405. The ISS program is a joint venture involving five space agencies. NASA, of USA, Russia's Roskosmos, Japan's JAXA, Canada's CSA, and ESA made up of agencies from France, Brazil, Malaysia, Italy, and South Korea.

406. There is strong evidence that Ganymede (one of

Jupiter's moons) has an underground ocean that is covered by a very thick layer of ice (about 800 km or 497 miles thick) and a rock shell.

407. The first time that Halley's Comet was recorded was in a Chinese chronicle.

408. A total solar eclipse can only last for 7.5 minutes.

409. All the pagers in the world stopped working when a satellite failed in 1998.

410. Makemake is essential in the history of the solar system as, along with Eris, it was one of the objects that led the International Astronomical Union to create a new classification group of dwarf planets.

411. In the southern hemisphere, you can find constellations such as Apus (bird of paradise), Chamaeleon (chameleon), Tucana (toucan), Lepus (rabbit), Grus (crane) and Hydrus (water snake).

412. The Milky Way's oldest star was formed just after Big Bang and is over 13.6 billion years old.

413. The Triangulum Galaxy actively creates stars that are scattered around the spiral arms. The rate of star birth is much higher than that of the Andromeda Galaxy.

414. Europa orbits Jupiter at a distance of 670,900 km (414,000 mi) from the planet. It takes 3.5 years to complete one orbit, and as it is tidally locked, the same sides always face Jupiter.

415. A nebula is usually made up of hydrogen and helium.

416. Earth's life cannot exist on Mars due to the low atmospheric pressure (amongst many other things like very little water).

417. Earth's core is about the same size as Mars, i.e., 7,091 km (4,400 mi) wide.

418. Some well-known nebulae are Eagle Nebula,

Barnard's Loop, Boomerang Nebula, Pelican Nebula, and Tarantula Nebula.

419. Phobos (one of Mars moons) has no atmosphere.

420. The tallest mountain on our moon, at 4,700 m (13,123 ft) high, is slightly over half of Mt Everest's height.

421. When the Milky Way and the Andromeda galaxy collide in about 2 to 4 billion years, it will form one large elliptical galaxy.

422. Galileo Galilei discovered Ganymede, one of Jupiter's moons, on 7 January 1610.

423. Eris is about the same size as Pluto, with a diameter of about 2,325 km (1,445 mi).

424. NASA's Pathfinder' small robot, Sojourner, was the first robot to explore a planet. That planet was Mars.

425. Edwin Hubble, one of the most important astronomers of the 20th century, discovered the shape and size of the Milky Way. He also proved there were other galaxies besides the Milky Way in a universe much bigger than our galaxy.

426. Pluto used to be the 9th planet from the sun, the smallest and furthest planet before it was demoted to dwarf planet status.

427. Mercury would damage the Hubble Space Telescope if it could view it due to its brightness from being so close to the Sun.

428. Earth means 'ground' or 'soil.'

429. The Whirlpool Galaxy and its companion, M51b, are connected by a bridge of dust and gas as they merge.

430. Black holes are full of debris that they have collected from space. They are not empty.

431. Spiral galaxies are listed as type S, with an a, b or c depending on the tightness of the spiral arms

and the size of the central bulge. A barred spiral galaxy with a long bar in the center has the symbol SB.

432. Earth's solar system is aged approximately 4.571 billion years.

433. Smoke, toxic gasses, dust, volcanic ash, and salt are all contaminants in Earth's atmosphere.

434. Satellites are programmed to avoid meteors, so they will not be hit and be destroyed.

435. Ganymede's (one of Jupiter's moons) temperature during the day ranges from -113 to -183 degrees C (-171 to -297degrees F).

436. A Type A spiral galaxy has the most tightly wound arms while the Type C spiral galaxy has very loose spiral arms.

437. The most violent weather in our solar system happens on Neptune.

438. In 1917 Thomas Wright first proposed the idea of a galaxy.

439. There have been over 2,400 research investigations in the ISS microgravity laboratory.

440. Space junk or space debris is any machinery or object left in space by humans. There are about 500,000 pieces of space junk in orbit that have the potential to destroy or damage other satellites in orbit.

441. Due to their size, the larger icy objects in the Kuiper Belt are known as dwarf planets. They are larger than asteroids and too small to be a planet.

442. Since its demotion to being a dwarf planet, Pluto now also has an asteroid number 134340.

443. The Milky Way and Andromeda Galaxies are getting closer together every day and will merge in about 5 billion years.

444. Nine spacecraft have launched missions to the

outer planets of our solar system. All of them have encountered Neptune.

445. The moon has an elliptical orbit around Earth.

446. Space has been growing bigger since the beginning of time.

447. On average, a lunar eclipse happens 2 to 3 times a year.

448. The Helix Nebula is the closest to Earth, approximately 700 light-years away! This means it would take you 700 light-years to travel there if you could travel at the speed of light!

449. The last supernova of four supernovae that have been found in the Pinwheel Galaxy was recorded in 2011.

450. Halley's Comet is known as a periodic comet, i.e., it orbits the sun in a period that is less than 200 years.

451. Makemake is the second brightest object in the Solar System, and Pluto is the brightest. You can see Makemake from your home high-end telescope as it is so bright.

452. 88% of Earth's iron is in the core.

453. The water vapor plumes detected on Europa, Jupiter's moon, in 2019 are similar to those observed on Enceladus, Saturn's moon.

454. The polar caps on Mars are mostly made up of frozen water and a thin layer of carbon dioxide. If the chilled water from its southern polar cap was melted, it could cover the whole planet with water about 11 meters or 36 feet deep.

455. When astronauts return from space, they often forget things fall due to Earth's gravity as they are used to objects floating around in space.

456. Galaxies often interact and can collide with each

other. When two galaxies collide, dust and gases flow, intermingle and form new stars.

457. The health of the atmosphere is significantly affected by humans. Damage to our atmosphere can occur due to the Greenhouse effect, global warming, destruction of the ozone belt, air contamination, and acid rain.

458. As of 2013, Jerry L Ross and Franklin Chang-Diaz, both Americans, have each been to space seven times.

459. Four spacecraft have flown past Saturn - Pioneer 11, Voyagers 1 and 2, and Casini-Huygens.

460. There has not yet been any evidence of life on Jupiter.

461. Black holes contain the same amount of mass as their original star.

462. No spacecraft has landed on Ganymede, one of Jupiter's moons.

463. It was only in 1833 that scientists believed that meteors came from our solar system.

464. Haumea was given the status of the dwarf planet as it had its gravity. However, it wasn't strong enough to remove similar objects from its region.

465. Because Mars' moons are not as large as Earth's moon and don't stabilize it, it tilts more towards the Sun, causing warmer summers than on Earth.

466. Earth is the largest inner planet, then Venus, Mars, and Mercury.

467. The Andromeda Galaxy is named after Andromeda, the mythological princess.

468. The naked eye can sometimes see the Triangulum Galaxy. However, it can usually be found with a pair of binoculars or a telescope.

469. The composition of Earth is mainly iron, oxygen, and silicon.

470. Earth's atmosphere comprises 78% nitrogen, 21% oxygen, and smaller amounts of other gases, including argon, carbon dioxide, helium, and neon.

471. Neptune's rings are not complete, so they are often called 'arcs.'

472. The Andromeda Galaxy is 260,000 light-years long.

473. The surface pressure of Jupiter and its high temperatures would make it impossible for any earth-life on the planet to exist.

474. A comet is material from our solar system that dates back to when the Sun and planets were formed. It is not a spacecraft or an alien base.

475. Venus is closer to the Sun than Earth and takes 225 days to orbit the Sun.

476. About 100 million pieces of junk orbit Earth at about 27,000 km/hr (17,000 mph).

477. The ancient civilizations did not find Neptune. It was only first seen in 1846 using mathematical predictions.

478. The Messier 87 Galaxy is not a spiral galaxy like the Milky Way. Instead, it is elliptical, and it is one of the most massive galaxies in our universe. Messier 87 is expanding as it continues to absorb smaller galaxies and matter.

479. The Sun is not round and is flat on the top and bottom.

480. The moon has three types of rocks, all found on Earth - basalt, anorthosite, and breccia.

481. Cosmonaut means 'universe sailor' and refers to an astronaut from Russia.

482. NASA has three spacecraft orbiting Mars, named Mars Reconnaissance Orbiter, Mars Odyssey, and MAVEN. ESA has two orbiting spacecraft called

ExoMars Trace Gas Orbiter and Mars Express. The 6th and last spacecraft to orbit Mars belongs to India and is named Mars Orbiter Mission (MOM).

483. Nicholas Copernicus thought that Mars moved backward as Earth, in its orbit around the Sun, overtook Mars.

484. Countries that are capable of launching satellites include the US, the UK, Russia, China, Israel, India, Ukraine, North and South Korea, France, Japan, and New Zealand.

485. The James Webb Space Telescope is scheduled to launch in 2021, replacing the Hubble Space Telescope. It is set to orbit Earth 1.5 million km (2.4 million miles) above Earth.

486. The Small Magellanic Cloud, 7.000 light-years, is half the size of the Large Magellanic Cloud, 14,000 light-years. In comparison, the Milky Way is much bigger, about 100,000 light-years.

487. Venus is Earth's closest neighbor and the second planet from the Sun.

488. The Hubble Space Telescope was scheduled to last until 2014, but it has continued to work to this day.

489. When many meteors come together, a meteor shower occurs.

490. It takes 27.3 days for our moon to orbit Earth.

491. The Milky Way Galaxy is smaller than the Andromeda Galaxy.

492. Titan, the largest Saturn moon, is made up mostly of water ice and rock, with a frozen surface of liquid methane and landscapes covered in nitrogen.

493. One galactic year is the time it takes for the Sun to orbit the Milky Way.

494. The Spanish telescope, Gran Telescopio Canarias is the largest reflecting telescope in the world.

495. As of 2013, Sergei Krikalev, a Russian, has been to space more than any other person. Including two ISS expeditions, he has traveled to space six times and spent more than 2.2 years (803 days, 9 hours, and 39 minutes) in space.

496. There has only been one spacecraft that has flown past Uranus.

497. Human sacrifices have been made to Venus by ancient cultures, such as the Skidi Pawnee Indians of North America.

498. The composition of the sun is 75% hydrogen and 28% helium.

499. The Asteroid Belt - where most asteroids orbit the Sun - sits between Mars and Jupiter.

500. Charon is named after Charon, the ferryman, a Greek mythological figure, who rows souls across the River Styx to Pluto's realm in the underworld. Charon means "fierce brightness."

501. Most asteroids that are on the path to colliding with Earth are destroyed when they reach our upper atmosphere.

502. In the TV show The Outer Limits, you can see black and white photographs of the Sombrero Galaxy in the credits at the end of each episode.

503. An artificial satellite is one that is sent to space to send information back to Earth.

504. Mark Twain was born on 30 November 1835 during an appearance of Halley's Comet, and he predicted he would die during the next one, which he did!

505. Driving at 96 kph (60 mph), you would reach Mars in 271 years and 221 days.

506. Johannes Kepler observed a supernova and

named it Kepler's Star in 1604, which faded after one year.

507. Jupiter's moon Ganymede is the largest in the solar system.

508. The Hubble Space Telescope discovered Uranus' two outer rings from 2003 to 2005.

509. A black hole is a very dense object in space where no light can escape.

510. The fourth-largest planet is Neptune.

511. There are more than one trillion stars in the Andromeda Galaxy.

512. You can see Haumea with a good quality telescope as it is very bright.

513. There are astronomical catalogs of nebulae such as the Barnard catalog, the Gum catalog, and the Sharpless catalog. They group astronomical objects by type, morphology, origin, means of detection, or method of discovery.

514. Eighteen people have died during four trips into space. Eleven others have been killed while training for spaceflight.

515. Eris is smaller than Earth's moon but could potentially fit all the objects in the asteroid belt in it.

516. When a star explodes and turns into a supernova, it will likely turn into a nebula and neutron star. However, a large explosion may result in a black hole!

517. Charon's (one of Pluto's moons) crust looks like it has been split open as it resembles a chasm four times as long as the Grand Canyon and twice as deep in some places.

518. Mars is the fourth planet from the Sun.

519. Some of the most well-known constellations are

Orion, Ursa Major, and Minor, Zodiac, and Pegasus.

520. Scientists believe that approximately three-quarters of the universe is made up of dark matter or energy.

521. As early as 460 to 370 B.C., Democritus became the first person to suggest that the Milky Way was made of stars.

522. In the center of the Pinwheel Galaxy, almost no stars are born.

523. The Soviet Yuri Gagarin was the first human in space in the spacecraft Vostok 1 in 1961, followed by American John Glenn Jr the same year. Neil Armstrong was the first person to land on the moon in 1969.

524. As there is no gravity on the International Space Station, astronauts can grow about 3% taller within six months. When they return to Earth, it takes a few months for them to shrink back to their original height.

525. Nebulae are enormous, often with diameters of millions of light-years.

526. Gas giants have different nicknames, for example, Giant Neptunes, Hot Jupiters, and Super Jupiters.

527. Neptune spins around very quickly on its axis.

528. The Voyager 1 space probe launched in 1977 will reach the Oort Cloud in about 300 years, taking another 30,000 years to travel through it.

529. When supernovas explode, elements such as gold and uranium are formed due to the high temperatures.

530. A spiral galaxy has a central bulge that contains a black hole.

531. The first telescopes used by merchants allowed

them to see upcoming trade ships, to gain a competitive advantage.

532. Twelve astronauts have walked on the moon. They were all from NASA's Apollo missions from 1969 to 1972.

533. Ceres is round, unlike many asteroids in the Asteroid Belt, which are irregularly shaped. It is large enough for gravity to shape it into a sphere.

534. The space station offers valuable opportunities for testing spacecraft systems and equipment and acts as a staging base for potential Moon or Mars missions.

535. During its orbit, Halley's Comet could be as close as Venus is to the sun or as far away as Pluto.

536. In 2007, when testing a missile, China shot down one of their satellites.

537. The planet with the most craters in the solar system is Mercury.

538. Depending on what meteors are made up of, they may have different colors.

539. We can successfully detect Jupiter's radiation from Earth.

540. From the vapor plumes that erupt from Ceres into space, astronomers believe that Ceres may have a subsurface ocean that could support life. Despite this, astronomers are more focused on exploring life on Europa and Mars, more than Ceres.

541. 17 and a half days was the longest space shuttle orbit.

542. A reflection nebula that is formed during a star's rapid evolution between the asymptotic giant branch phase and the subsequent planetary nebula phase is called a protoplanetary nebula. The Red Rectangle Nebula is an example of a protoplanetary nebula.

543. Jupiter weighs double the total weight of all the other planets put together.

544. The same person who discovered Haumea discovered its two moons.

545. When there are no bright lights around, you may be able to see the Andromeda Galaxy! It is the furthest object you can see with the naked eye.

546. Neptune's atmosphere comprises mainly hydrogen and helium and some methane.

547. Most asteroids have irregular shapes and are too small to be affected by Earth's gravity.

548. Venus has no seasons as it doesn't tilt on its axis.

549. It can take about an hour for total daylight to come back after a total solar eclipse.

550. In 1922, the celestial sphere was divided into 88 constellations by The International Astronomical Union and an American astronomer Henry Norris Russell.

551. Jupiter's intense magnetic field produces exceptionally high radiation levels on its moon, Europa, strong enough to kill a person in one day.

552. The universe is so big that we can only see about 5% of the universe.

553. Uranus has two sets of rings around it. Nine inner rings are narrow and dark grey. It has two dusty rings, and two outer rings, one reddish in color, and the outermost one is blue. The rings contain dust-sized particles to large boulders.

554. The sun is our solar system's star.

555. When a total solar eclipse happens, the temperature of the air will drop suddenly by about 20 degrees, and the immediate area turns dark.

556. Astronomers use the Andromeda Galaxy to

understand the evolution and origin of galaxies as it is the nearest spiral galaxy to Earth.

557. Mercury has extreme temperatures. The side that faces the Sun can have a temperature of up to 427 deg C (800 deg F) while the opposite side away from the Sun can be as cold as -173 deg C (-279 deg F). The reason for this is that Mercury doesn't have an atmosphere to regulate its temperatures.

558. Mars and Earth have similar seasons as they have similar tilts.

559. Earth is the only planet that is not named after a god.

560. Time slows as you approach a black hole.

561. Occasionally there is no full moon in February.

562. The ISS was designed to be a laboratory and observatory for space environment research. Crew members could conduct experiments in many scientific fields, including biology, human biology, physics, astronomy, and meteorology.

563. There is a six-sided jet stream coming out of Saturn's north pole.

564. There is wi-fi on the moon, beamed up from Earth with four infrared telescopes!

565. Ceres is the largest asteroid with a diameter of 933 km. It is also a dwarf planet.

566. Mars has been given many names in the past by different civilizations. The ancient Babylonians called it Nergal, which means 'Star of death.' The Hebrews called it Ma'adim, meaning 'One who blushes' and the Greeks and Romans named it after their gods of war.

567. Two men from each of the six different Apollo missions have walked on the moon, including Neil Armstrong and Buzz Aldrin.

568. The moon is not round - it is oval like an egg.

569. The Hubble Space Telescope has the same storage capacity as 20 car batteries.

570. Meteoroids that fly into Earth's atmosphere and then bounce out again are called Earth grazing fireballs.

571. Yerkes Observatory in Wisconsin has the largest refracting telescope. It was built in 1897.

572. Pioneer 11 in 1979 discovered Saturn's two outer rings.

573. Neptune completed its first 165-year orbit around the sun in 2011 since it was first discovered in 1846.

574. Different names were suggested for Uranus in the past. These include Hypercronius meaning above Saturn, Georgium Sidus, meaning the Georgian Planet and after King of England, George III.

575. There are six faint rings around Neptune.

576. The Milky Way has two major arms spiraling out from the center bar, not four, as initially thought. The names given to these two arms are Scutum-Centaurus and Perseus.

577. Charon, Pluto's largest moon, is the largest moon in comparison to its world than any other moon in the solar system. The second-largest moon, in contrast to its planet, is Earth's moon.

578. The sun takes 225 - 250 million years to orbit the Milky Way.

579. In the PC game Descent, one of the secret levels takes place on Ceres.

580. Saturn's moon Titan receives only 1% as much sunlight as Earth, although 90% of the sunlight is absorbed by the thick atmosphere.

581. Richard Assmann and Leon Teisserenc de Bort are considered early pioneers of aerology (the study of the atmosphere).

582. Very little light reaches Venus due to its thick clouds. From space, it looks white.

583. One day the sun will become the same size as Earth.

584. Astronomers believe the center of all galaxies has a black hole.

585. The Hubble Space Telescope took 50 years to be developed, built, and launched.

586. There are several hypotheses about how Phobos and Deimos, Mars two moons, came about. Some scientists think that they came from the asteroid belt and Jupiter's gravity pushed them into orbit around Mars. A second hypothesis is that they were formed by dust and rock that was drawn together by gravity as satellites around Mars. A third hypothesis is that Mars collided with an existing moon, which then formed Phobos and Deimos from the dust and rubble out of the collision.

587. When an asteroid does hit Earth, a crater can be formed.

588. Scientists have observed signs of water trickling down Mars' crater walls and cliffs, which indicate that there may be water on Mars. As the water is not frozen, it is likely to be salty.

589. It takes Jupiter 12 Earth years to orbit the Sun.

590. The temperature on Venus is boiling and can go up to 471 degrees C (879 degrees F).

591. The moon will be about 23,500 km (14,600 mi) further away from Earth in about 500 million years.

592. Al Tarif, Beta Cancri, is the brightest star in the Cancer constellation.

593. There is no friction or gravity in outer space so that planets can orbit the sun.

594. Asteroids often collide with each other, which ends up throwing them out of orbit or hitting other planets.
595. Scientists believe that beneath Uranus' hydrogen methane atmosphere is a hot ocean of water, ammonia, and methane, which covers a rocky core. It has no solid surface.
596. The Sun takes 225 to 250 million years to orbit the Milky Way.
597. Ceres is the largest body in the asteroid belt.
598. After the Sun and the moon, the second brightest object in the night sky is Venus.
599. The largest dust storms in the solar system are on Mars.
600. People are now more aware of asteroids after the Shoemaker-Levy comet collided with Jupiter in 1994. Different Hollywood movies, such as Deep Impact and Armageddon, have also increased public awareness.
601. It is believed that the universe has more stars in it than Earth has sand at its beaches.
602. Earth has five atmospheric layers. In order from the closest to Earth are the troposphere, stratosphere, ozone layer, mesosphere, and thermosphere.
603. The southern hemisphere on Mars, with lots of craters, is very different from the northern hemisphere with fewer craters.
604. On Io, one of Jupiter's moons, there are over 400 active volcanoes.
605. Jupiter has sixty-seven known moons. Several of those moons were named after the many lovers of the Roman god Jupiter.
606. Earth only has one moon.
607. Before Eris was reclassified as a dwarf planet in

2006, it was considered to be the tenth planet (after Pluto, who also lost status as a planet at the same time).

608. The Cancer constellation is very difficult to see as it is one of the dimmest constellations.

609. Mars has six spacecraft in orbit, and two robots are roving its surface. NASA and ESA have plans to send further robots to Mars.

610. The core of Earth makes up about 30% of its mass, whereas the center of the moon makes up for 2-4% of its mass.

611. Chile is currently building two reflecting telescopes that will be bigger than the biggest telescope of today. The Giant Magellan Telescope will have a mirror with a diameter of 24.4 meters (80 ft), and the European Extremely Large Telescope will have a mirror with a diameter of 39.9 m (129 ft). In comparison, the 30 Meter Telescope, currently being built in Hawaii, will have a diameter of 30 m (98 ft). The Gran Telescopio Canarias (the biggest telescope as of 2019) only has a mirror with a diameter of 9 meters (30 ft).

612. Eris is slightly larger than Pluto but smaller than Earth's moon. It was discovered in 2003 and took 557 years to orbit the Sun.

613. In ancient times, astrology was used to determine the timing of specific cultural celebrations.

614. Before Ceres was classified as a dwarf planet in 2006, it was considered to be a planet and then an asteroid.

615. Although Callisto (one of Jupiter's moons) is about twice the distance from Earth than the moon, when viewed through a telescope, it is much

brighter than our moon due to Sun's reflection of its ice layer.

616. X-ray astronomy is the use of X-ray wavelength to study space objects.

617. More comets impact Jupiter than any other planet in our solar system.

618. The word telescope means "far" and "to look and see" in Greek.

619. Triton keeps the same face towards Neptune as it rotates on its axis and orbits the planet.

620. The Triangulum Galaxy is considered a satellite of the Andromeda Galaxy because of its proximity to one another.

621. Triton is approximately 354,800 km (220,405 mi) from Neptune.

622. Our solar system makes up less than one-trillionth of our universe.

623. Pluto is symbolized by a P interlocked with L. These happen to be the first two letters of Pluto and also Percival Lovell's initials. Percival was the astronomer at the Lowell Observatory, also named after him, who started to search for planets beyond Neptune.

624. A star only looks like it twinkles due to turbulences in the Earth's atmosphere. It actually doesn't.

625. Mars' atmosphere mainly contains carbon dioxide.

626. Scientists estimated that in 3.6 billion years, Neptune's largest moon Triton would be torn apart.

627. Two types of terrain exist on the moon. The highlands are the bright terrain as they're higher than 'maria.' The 'maria' have a lower elevation and make up the dark terrain.

628. Pluto was considered the furthest and ninth planet

from the Sun for 76 years until it was demoted to being a dwarf planet in 2006.

629. The Whirlpool Galaxy has a massive black hole at the center of its spiral.

630. Jupiter turns on its axis every 9 hours and 55 minutes. This rotation flattens the plan and gives it an oblate shape.

631. The IAU demoted Pluto from being a planet to a dwarf planet in 2006.

632. Jupiter has 79 moons orbiting it, currently the largest number of confirmed moons surrounding any planet in our solar system. The four largest, Ganymede, Callisto, Io, and Europa, all named after Galileo Galilee, are called the Galilean Moons. Saturn is waiting for some newly discovered moons to be confirmed, and it will then become the planet with the most moons.

633. Ceres does not have a moon.

634. The Russian satellite Sputnik was the first human-made object that went into space.

635. An earth observation satellite is used to make maps and observe environmental changes.

636. Although it is further away from the sun, Pluto's orbit took it closer than Neptune to the sun in 1979.

637. Scientists don't expect to find any life on Mars, but they are looking for a life that may have existed on Mars a long time ago when it may have been covered with water.

638. Hyakutake was a long-period comet that was discovered in January 1996 by a Japanese amateur astronomer, Yuji Hyakutake. Later the same year, the Ulysses spacecraft accidentally traveled through Hyakutake's long tail, which was about half a billion km (more than 300 million miles)

long. Hyakutake was a comet from the Oort Cloud.

639. If you boil water in space, it creates one giant undulating bubble (as compared with lots of little bubbles in boiling water on Earth).

640. As of December 2019, Saturn has 53 confirmed moons and 29 awaiting more information so they can be verified. If all of them are approved, Saturn will have the largest number of moons relating to one planet in our solar system.

641. There is no twilight before nightfall as there's no atmosphere on the moon.

642. It is said that Halley's Comet started in the Oort Cloud.

643. Outer Space was first coined in a poem The Maiden of Moscow by Lady Emmeline Stuart-Wortley in 1842.

644. Ceres features in the TV series The Expanse where humans inhabit Ceres.

645. Clyde Tombaugh, an American astronomer, discovered Pluto in 1930, but Charon, Pluto's moon, wasn't discovered until 1978.

646. Every year, approximately 500 meteorites hit Earth. However, only a handful ever make it to scientists to study as most of them fall into the ocean.

647. Cosmonauts and astronauts have conducted more than 205 spacewalks aboard the ISS since 1998. Spacewalks are performed for maintenance and repair, as well as the construction of space stations.

648. The first artificial satellite to be launched into space was the Soviet Union's Sputnik 1. It was launched in October 1957.

649. Valentina Tereshkova, a Russian, was the first

woman in space. In 1963 she orbited Earth for nearly three days on the spaceship Vostok 6.

650. Proxima Centauri is the closest neighbor to Earth. It is part of the Alpha Centauri cluster of stars and 4.3 light-years away. A spacecraft would take 25,000 years to reach Proxima Centauri.

651. Makemake, a dwarf planet, is so far away from the Sun, making it very cold with an average temperature of -243.2 degrees C (-405.7 deg F). Scientists think methane, ethane and nitrogen ices cover its surface.

652. The dwarf planet Sedna, discovered in 2003, is believed to be located in the inner Oort Cloud.

653. A galaxy with no spiral, lenticular, or elliptical structure is called an irregular galaxy. The large and small Magellanic clouds that border the Milky way has no distinctive shape as they are within the gravitational force of other nearby galaxies.

654. Water erosion on Mars may have created its channels and canyons.

655. The Magellanic Clouds differ from the Milky Way in several ways. Not only do they have a different structure and a lower mass, but they also are rich in hydrogen and helium gases, and they have tiny metals. In comparison, the Milky Way is 70% hydrogen, and the rest is carbon dioxide, ammonia, and formaldehyde.

656. A day on Venus lasts 243 Earth days.

657. Jupiter is the solar system's fourth brightest object.

658. Where Earth's crust is made up of several moving plates, Mars' crust is only made of one piece and is therefore much thicker than Earth's.

659. The largest black hole that has been found in the center of a galaxy is in the Sombrero Galaxy.

660. The thermosphere, the furthest atmospheric layer

to Earth, contains the ionosphere and exosphere and about 0.001% of all gases in the atmosphere. It can extend out as far as 9,656 km (6,000 miles) into space.

661. Ceres takes 4.6 Earth years to orbit the Sun, traveling about 413,700,000 km (257,061,262 mi). It takes 9 hours and 4 minutes to rotate around its axis.

662. The largest crater on the moon that we can see from Earth, with a diameter of 294 km (183 mi), is the Bailly Crater.

663. Voyager discovered the first 9 of Saturn's moons.

664. Big supergiant and hypergiant stars burn up faster than smaller stars and explode into bright supernovae when they die.

665. The names of Uranus' rings, from the center, are Zeta, 6, 5, 4, Alpha, Beta, Eta, Gamma, Delta, Lambda, Epsilon, Nu and Mu.

666. The red dwarfs are the most common stars. They live longer and shine less than any other type of star. Our sun, in comparison, is a yellow dwarf star.

667. Going to the center of our Milky Way galaxy would take longer than humans have existed on Earth!

668. Due to its thick atmosphere, Venus is the hottest planet even though Mercury is closer to the Sun. The dense toxic environment traps the heat on Venus.

669. Crews aboard the ISS have eaten more than 25,000 meals since 2000.

670. The Hubble Space Telescope is the first telescope that was designed to be repaired in space. Astronauts have fixed it in space five times. The last repair mission was in 2009.

671. A biosatellite is one used to carry living organisms for scientific experiments.

672. At one time, it was thought that comets only passed through the solar system once. Edmond Halley was the first person to discover that comets could have a periodic orbit, and in 1705 he suggested that the comets seen in 1531, 1607, and 1682 were the same Comet. He predicted the next time we would see Halley's Comet was in 1758-1759. He was right but died 16 years before the comet came around again! Halley's Comet was named in his honor.

673. Only two spiral galaxies can be seen with the naked eye from the southern hemisphere - Andromeda galaxy and the Milky Way.

674. The Big Bang Theory, the Lambda-CDM model, and dark matter are all theories of theoretical astronomy.

675. The Triangulum Galaxy received its name from the constellation Triangulum.

676. Constellations that are located in the same area of the sky are sometimes grouped and given a family name after the main constellation in that area.

677. The Local Group is a cluster of about 30 galaxies.

678. In the northern hemisphere, you can find constellations such as Lyra (harp), Orion (hunter), Aquila (eagle), Bootes (Herdman), Perseus (Medusa's killer), Ursa Major (big bear) and Ursa Minor (little bear), Draco (dragon), Andromeda (princess) and Canis Major and Minor (big and small dog).

679. Asteroid means 'star-like' and was given its name in 1802 by William Hershel.

680. An astronaut's footprint will last on the moon for

millions of years as there is no weather or atmosphere on the moon.

681. Ceres can be found in the Asteroid Belt, which is home to many asteroids of different sizes and shapes.

682. It is believed that the Solar System formed about 4,568 years ago.

683. A solar eclipse will usually alternate with a lunar eclipse.

684. The space shuttle used oxygen and hydrogen as fuel.

685. A space shuttle could only be launched in perfect weather conditions.

686. The dwarf planets, Pluto, Haumea, and Makemake, are all situated within the Kuiper Belt.

687. A weather satellite monitors the weather.

688. There is no evidence of strong radio or x-ray emissions from the center of the Pinwheel Galaxy, which indicates it doesn't have a black hole.

689. In 1930, Pluto was named by an 11-year-old girl, Venetia Burney, who suggested the name as it was so dark and like the god of the underworld. She was given a 5-pound note as a reward for choosing the name.

690. Quasars, the furthest known objects in our universe, are a matter which breaks apart as it goes into a black hole. The nearest Quasar is billions of light-years away.

691. The mass of Mercury,330,104,000,000,000 kg (727,754,745,962,768.25 lb) is 0.055 times Earth's mass.

692. The Hubble Space Telescope can see more than the naked eye - it can also see ultraviolet and near-infrared.

693. A plutoid is also called an ice dwarf.

694. Ceres was named after the Roman Goddess of corn and harvests. The word 'cereal' comes from the same name.

695. The center of the black hole is the Singularity, which has the strongest gravitational pull.

696. Eris is the largest dwarf planet, 27% larger than Pluto, which comes next.

697. Low gravity in space makes it hard for an astronaut to tell if they need to urinate, so they go to the toilet to empty their bladder every 2 hours.

698. The sun supports all life on Earth through photosynthesis.

699. Triton (one of Neptune's moons) is the largest moon in our solar system. With a diameter of 2,706 km (1,681 mi), it is the 16th largest object and bigger than Pluto or Eris.

700. The gravity on Mars is about 37% of the Earth's so you could jump three times higher on Mars.

701. To the Greeks, the Milky Way was known as the Milky Circle. They believed that the goddess Hera was surprised to find another woman suckling Hercules while she slept.

702. Pluto was reclassified as a dwarf planet when Eris was discovered.

703. NASA hopes a human-crewed mission to Callisto may be possible in the 2040s.

704. About 70% of Earth is covered in water.

705. The average distance between Callisto and its planet Jupiter is 1,882,700 kilometers (1,169,856 miles).

706. The circumference of Uranus at its middle is 159,354 km (99,017 mi).

707. Walt Disney created a cartoon dog called Pluto in 1930 and named it after the planet (it was a planet at the time!)

708. Uranus' thick atmosphere gets extremely dense the deeper you go.

709. Uranus has 27 moons in orbit.

710. It was possible for travelers to pay for their passage into space at one time, but was stopped in 2011 when the ISS crew was reduced to 6.

711. When standing on Mars, our Earth sunset looks blue.

712. In 1972, the last person to step on the moon was Eugene Cernan.

713. The Great Red Spot is a massive storm on Jupiter. This storm has raged for at least 350 years and is so large that three Earths could fit inside it.

714. The five officially recognized dwarf planets are Ceres, Pluto, Haumea, Makemake, and Eris, in order from their distance from the Sun.

715. Some people believe that the Wise Men saw Halley's Comet and not the Star of Bethlehem when Jesus was born.

716. Once the ISS reaches its end of use, it will use many modules for other purposes and space stations.

717. Mt Chimborazo, a mountain in the Andes with a height of over 6,096 meters (20,000 feet), sits higher on Earth's bulge and is, therefore, the closest point to space, not Mt Everest which is only the tallest mountain from sea level.

718. Haumea rotates very quickly on its axis, which means a day on Haumea lasts about 3.9 hours.

719. Most meteors burn up in our atmosphere before it lands.

720. The windiest planet in our solar system is Neptune.

721. Neptune has thin wispy type clouds that cover the planet.

722. Phobos is moving closer to Mars by about 2 meters (6.5 ft) every 100 years, so in nearly 30 - 50 million years, it will collide with Mars.

723. There is debate about whether life existed on Mars from a Martian meteor that hit Earth 13,000 years ago in Antarctica. Scientists believe the meteor had microscopic fossils of bacteria indicating there may have been life once on Mars.

724. Babylonian astronomers first recorded Venus in the 17th century BC.

725. The sun makes up 99.86% of the Solar System while Jupiter and Saturn make up most of the rest.

726. Without gravity in space, nothing pushes the bubbles up in fizzy drinks in space. You won't be able to burp out the gas from Coca Cola!

727. Scientists believe the universe is flat, even though the planets aren't.

728. The Polynesians have relied on the Magellanic Galaxies to predict wind as well as for navigation.

729. Mercury has a large metallic core, partly molten and partly liquid.

730. M51b, the Whirlpool Galaxy's companion, is a dwarf galaxy.

731. About 100 satellites are launched every year into space.

732. One million Earths could fit inside the Sun, which is considered to be an average-sized star.

733. NASA is working on proving warp drives and travel faster than the speed of light. The space around a spacecraft would need to move, not the spacecraft itself.

734. 80% ore more of Uranus' mass is made up of a hot dense fluid of "icy" materials—water,

methane, and ammonia—above a small rocky core.

735. A high-speed wind on Venus carries its clouds every four days around the planet.

736. Europa, with a diameter of 3,100 km (1,900 mi), is larger than Pluto and smaller than Earth's moon.

737. Galaxies sometimes merge to form a bigger one, or they destroy each other.

738. A black hole absorbs light and everything else in space.

739. In 1964 the first black hole Cygnus X-1 was discovered.

740. The fuzzy outline of a comet (also called a coma) is created when it gets close to the Sun.

741. Astronomers do not know how cosmic rays are created.

742. The diameter of the Pinwheel Galaxy is approximately 170,000 light-years.

743. Jupiter is the fifth planet from our Sun and the largest planet in the solar system with a radius of nearly 11 times the Earth's size.

744. The mass of Uranus is 86,810,300,000,000,000 billion kg which is equivalent to 14,536 times Earth's mass.

745. Uranus is the smallest of the giant planets.

746. Charon was first called S/1978 P 1 after the year it was discovered and is the first object around Pluto.

747. Haumea is the most recent dwarf planet to be named. There are only five dwarf planets.

748. SN 1006 was recorded as the brightest supernova in 1006 AD.

749. To find their way around the sky, astronomers use

the constellations, such as Orion, the zodiac signs, and Ursa Major, to guide them.

750. The Dutch astronomer Christiaan Huygens discovered Titan, a Saturn moon, in 1655.

751. The Andromeda Galaxy is the closest galaxy to the Milky Way and is about 80,000 light-years away.

752. The first black hole to be imaged sits in the middle of the M87 Galaxy.

753. For a star or an object to leave the Milky Way, it must have enormous amounts of energy and speed.

754. Earth's powerful magnetic field protects the planet from the effects of the solar wind.

755. The majority of meteors are the size of a pebble.

756. The temperature inside the sun is about 15 million deg C (27 million deg F).

757. Earth's orbit slows down about two milliseconds every 100 years.

758. A meteoroid is also known as a 'space rock,' i.e., only in space. When it enters the Earth's atmosphere, it is called a meteor.

759. Astronomers believe that an asteroid of 15 km (9.3 mi) diameter exploded over Siberia and caused damage of a radius of hundreds of kilometers/ miles.

760. A comet's nucleus is made of ice and can range from a few meters (feet) to a few kilometers (miles) wide.

761. One Neptune year is equivalent to 165 Earth years.

762. The Whirlpool Galaxy is a spiral galaxy. William Parsons discovered its helical structure in 1845, looking through his telescope in Ireland.

763. Keo is the name of a space-time capsule designed to carry messages from the present Earth to humans in 50,000 years. It was supposed to have launched in 2003 and still hasn't been launched in 2019.

764. Some astronomers believe that objects sucked into black holes can come out in another galaxy.

765. The ancient Romans and Greeks name the moon's phases after three goddesses. In essence, Artemis (Diana) for the new moon, Selene, for the full moon, and Hecate was the dark side of the moon.

766. The surface of the sun is 11,990 times bigger than Earth's.

767. Galaxies that form a lot of new stars at a fast rate is called a starburst.

768. If you were on Pluto, you could see stars during the day as the sky is so dark.

769. It has been suggested that more than one supermassive black hole exists in the M87 Galaxy.

770. It is not very easy to tell the difference between a meteorite and merely a rock. Dark meteorites can be seen easily in sandy deserts and equally in icy regions like Antarctica.

771. A constellation begins at dusk in the east and ends at dawn in the west.

772. Martian seasons are extreme because of their elliptical orbit around the Sun. When it is closest to the Sun, its southern hemisphere experiences a brief scorching summer. In contrast, its northern hemisphere goes through a cold winter at the same time. Then at its furthest point from the Sun, the northern hemisphere has a long mild summer while the southern hemisphere goes through a long cold winter.

773. Similar to Saturn's moon, Enceladus, Eris is

thought to be one of the most reflective dwarf planets in the solar system.

774. Different astronomers have different hypotheses about what will happen to our universe. The first is that it will eventually collapse into something else. The second is that it will keep expanding and growing forever until everything is so far apart that the universe will die. The third is that we have a flat universe that will stay the same and continue indefinitely. The last scenario is known as 'the Goldilocks effect' as everything is 'just right.'

775. A refracting telescope uses a concave and a convex lens.

776. Dark matter hasn't been measured, but scientists believe it is the glue that holds the universe together.

777. We can only see about 10% of the Milky Way as dark matter makes up the remaining 90%.

778. Meteoroids can travel through space at speeds of up to 42 km per sec (26 miles per sec).

779. The ISS travels at an average speed of 27,724 kilometers (17,227 mi) an hour.

780. A Mars day is equivalent to 24 hours 37 minutes on Earth.

781. Fireball meteors are hard to see as they often happen during the day and over the ocean.

782. Astronomers make assumptions about the Milky Way based on their observations about the Andromeda Galaxy. This is because both galaxies are very similar in nature.

783. 3753 Cruithne has its orbit around the Sun, but it looks like its following Earth's orbit. It is 5 km (3.1 mi) wide and is occasionally called our second Moon.

784. Pluto was the first object to be discovered in the Kuiper Belt.

785. Jupiter's magnetic field is 20,000 times stronger than Earth's magnetic field.

786. Mars' orbit around the Sun is 687 Earth days, which is twice as long as Earth's orbit of 365 days. As a result, Mars' seasons are twice as long.

787. Eris, a dwarf planet, is so far away from the sun that its atmosphere continually freezes and collapses on the planet. Scientists believe that the sun will begin to thaw Eris' icy surface as it moves closer to the sun. They believe there is a rocky surface under the ice.

788. The Messier 87 Galaxy is made up of gas and dust, surrounded by hot gas.

789. The chances of a rocket or spacecraft colliding with an asteroid are about one in a billion.

790. The Spanish Gran Telescopio Canarias telescope contains a mirror with a diameter of over 9 meters (30 ft).

791. NASA has nicknamed two satellites that are chasing each other in space, Tom and Jerry.

792. As of 2013, there have been 38 expeditions to the ISS. An expedition may last a maximum of six months. Early expeditions had crews of 3 people. This was reduced to teams of 2 for safety. However, the crew numbers now regularly reach six people.

793. An emission nebula, for example, the Omega Nebula, is made up of ionized gases that emit light of various wavelengths and different colors. The cause of an emission nebula is usually by a nearby hot star.

794. As far as we know, Mars has the longest valley in our solar system. Valles Marineris is more than ten

times longer than the Grand Canyon at a length of 4,000 km (2,500 mi).

795. The only dwarf planet with an atmosphere is Pluto. It is too thin and poisonous for humans. When Pluto is closest to the sun (at its perihelion), its atmosphere is gas. When it's furthest from the sun (at its aphelion), the atmosphere turns into ice.

796. In 1969 NASA's Apollo 11 was the first human-crewed spacecraft to land on the moon.

797. The sun is the closest star to Earth.

798. The first telescopes invented were all refracting telescopes and used glass lenses.

799. The Milky Way is only one of the millions of galaxies in the universe.

800. Venus' surface has been described as similar to molten lead.

801. The solar system orbits the Milky Way galaxy at a speed of about 220 km/sec or 136 miles/sec, which is approximately 0.073% of the speed of light.

802. New stars are being formed at the center of the Whirlpool Galaxy at a fast rate as a result of its interaction with its companion M51. This is expected to last no more than another 100 million years.

803. The first Kuiper Belt Object (KPO) was discovered in 1992 by Dave Jewitt and Jane Luu. It was a reddish colored speck further away than Pluto, and they wanted to call it Smiley! It has since been named 1992 QB1 (boring!).

804. The spiral arms of the Milky Way contain the new stars, whereas its center has mainly old stars.

805. There are more than 200 billion stars in the universe.

806. The Messier 87 elliptical Galaxy is located in the constellation Virgo.

807. A meteor shower occurs when a planet such as Earth moves through a stream of material in the comet's orbit.

808. Mercury is named after the Roman messenger to the gods.

809. The Cancer constellation can be seen in the southern hemisphere from summer to autumn.

810. Hydrogen is the main constituent of Saturn.

811. Cancer is the star sign for people born between 22 June and 22 July. Cancerians are said to be devoted to their family and their home and can be 'crabby' sometimes.

812. Amateurs can find the Sombrero Galaxy easily with a good set of binoculars or a good telescope halfway between Virgo and Corvus.

813. Scientists have identified over 1,000 objects in the Kuiper Belt with many hundreds of thousands more to discover.

814. Callisto, one of Jupiter's moons, may have a subsurface ocean where life could exist.

815. Earth's core has enough gold to cover the entire surface of the planet.

816. In 2004, astronomers believed a meteorite that fell to Earth in Yemen in 1980, may have come from Phobos, a Mars moon.

817. No one has been to the moon since 1972. This is because the Apollo missions cost much money, and there was more about America showing more power than the Soviets during the Cold War. When the Cold War ended, moon expeditions were not given the same budget as previously. In 2010 President Obama canceled the Constellation program, which George Bush started. Obama

directed funds be given to send astronauts to an asteroid that was closer to Earth, but this was stopped by President Donald Trump in December 2017 when he ordered NASA to send astronauts back to the moon.

818. Phobos, one of Mars' 2 moons, is not round and is one of the darkest objects in our solar system.

819. Mars is much lighter than Earth, and it is the last rocky planet from the Sun; all other planets after Mars are gas planets.

820. All moons have two things in common - they both orbit a planet and reflect light from the sun.

821. In Cherokee legend, the Milky way is called 'the way the dog ran away' after a dog who stole some cornmeal and was chased away.

822. The 13 zodiac constellations have the same names used for astrology plus Ophiuchus.

823. On 15 December 1970, the Russian spacecraft, Venera 7, became the first successful spacecraft to land on Venus.

824. About 21 million light-years from Earth sits the Pinwheel Galaxy.

825. Scientists once thought that Earth was the center of the Universe until Copernicus proposed that the Sun was the center. Due to the Universe expanding equally at all places, there is no center!

826. Approximately 20% of the air freezes on Mars during winter.

827. Scientists believe that the moon once belonged to Earth, and when Earth collided with a large object, a piece broke off and became our moon.

828. Charon, Pluto's largest moon, is covered with mountains, canyons, and landslides. It has a unique large mountain called 'Mountain in a Moat' coming out of a depression.

829. Long-period comets come from the Oort Cloud caused when something happens to one of the icy objects in the Oort Cloud, resulting in it falling towards the sun. Comets C/2012 S1 (ISON) and C/2013 A1 Siding Spring are two examples of Oort Cloud comets.

830. The Sombrero Galaxy is a lenticular (lens-shaped) galaxy in the Virgo constellation.

831. In the 1600s, astronomers once thought that the Andromeda Galaxy was part of the Milky Way.

832. Relative to its planet, our moon is the largest in the solar system.

833. A total of 4 galaxies can be seen with the naked eye - the Small and Large Magellanic Clouds, the Milky Way and the Andromeda Galaxy.

834. NASA's Voyager 2 spacecraft, launched on 20 August 1977, flew past Uranus in 1986, about 81,500 km (50,641 mi) from the planet.

835. Scientists estimate about 200 dwarf planets are waiting to be discovered in the Kuiper belt and more than 10,000 outside of the Kuiper belt.

836. A supernova explosion results in star matter being blown away to form new nebulae, which in turn makes new stars.

837. The sky is blue because the atmosphere absorbs the real color, purple.

838. The Kuiper Belt contains frozen gases, including methane, ammonia, nitrogen, and water.

839. Earth's density varies in different parts; for example, the crust is less dense than its metallic core. Earth's density averages 5.5 grams per cubic centimeter.

840. Inner planets have volcanoes, canyons, craters, and mountains on their surface. Earth is the only planet known to have water on its surface.

Other planets have underground or subsurface oceans.

841. The International Space Station (ISS) has two bathrooms, a gym, and is bigger than a six-bedroom house.

842. Our ocean tides are caused by the gravity from our moon and the Sun.

843. It is unlikely for a meteor to fall on a human being as they mostly fall into the ocean.

844. The Oort Cloud is made up of hundreds of billions or trillions of icy objects.

845. NASA mapping satellites discovered Mayan ruins, which were overgrown by jungle and may never have been found.

846. Mercury is the smallest planet in our solar system. A Mercury day lasts 176 Earth days.

847. The circumference of Mars is 21,297 km (13,233 mi).

848. The Kuiper Belt is thought to have been created at the same time as our solar system.

849. Sagittarius A* is the name of the black hole in the center of the Milky Way.

850. NASA and ESA are working together to plan a human trip to Mars by 2035.

851. Never look directly at a total solar eclipse as it can make you blind.

852. Our solar system consists of the Sun, planets and dwarf planets, moons, asteroids and comets, and stars.

853. Eris is about one fifth the size of the Earth's radius with a radius of 1,163 km (722 mi).

854. The very bright stars you can see in spiral galaxies are new large stars.

855. The largest known galaxies are the giant elliptical-shaped galaxies.

856. Before its launch, the Spitzer Space Telescope was named the Space Infrared Telescope Facility.

857. Thomas Harriot was the first person to draw a map of the moon as it looks through a telescope.

858. A NASA flight controller, John Aaron, saved Apollo 12 when it was struck by lightning. He also developed a safe way for Apollo 13 to come back to Earth.

859. Mars has a symbol that resembles a shield and a spear from the God of war. The male gender has the same symbol.

860. The New Horizons spacecraft that was launched in 2006 finally flew past Pluto on 14 July 2015, making the first close up observation of a Kuiper Belt object.

861. The space shuttle launched 135 missions in its 30 years from 1981 to 2011.

862. Refractor telescopes should only be under 1 meter (40 in) long to make it easy to use.

863. The South Pole-Aitken, the largest crater on the moon, is also the largest crater in our solar system. It has a diameter of 2,500 km (1,550 mi) and is located on the far side of the moon.

864. In 2013, Felix Baumgartner, skydiver, jumped from the stratosphere 36,576 meters (120,000 feet) above Earth.

865. The first dwarf planet to be visited by a spacecraft is Ceres in 2015. The spacecraft was called Dawn and was not manned. On the same trip, Dawn also surveyed a protoplanet Vesta.

866. The stars that we see in the sky are about 40 million km (25 million mi) away.

867. In a million years, you won't be able to see a solar eclipse as the moon is moving further and further away from Earth.

868. The Sun appears half the size it does on Earth when you are on Mars.

869. The flattest planet in our solar system is Saturn.

870. After Jupiter (63 moons) and Saturn (61 moons), Uranus has the most number of moons (27 moons).

871. The mesosphere is the fourth closest atmospheric layer to Earth and is about 50 to 70 km (31 to 53.5 mi) from the Earth's surface.

872. Ceres was discovered by an Italian astronomer Giuseppe Piazzi on 1 January 1801, who was searching for a star.

873. Radio astronomy is the study of radiation with wavelengths of 1mm or more.

874. Callisto takes 16.7 days to orbit Jupiter, a distance of 11.2 million km (7 million miles).

875. The Hubble Space Telescope is about the size of a school bus.

876. Clyde W Tombaugh discovered Pluto on 18 February 1930. It was the ninth planet from the Sun.

877. Saturn is known best for its rings, which are made up of ice and small amounts of dust. Saturn's rings are thought to be made of comets or asteroids that were torn apart from its strong gravitational pull.

878. Depending on the temperature of the star, it can be brown, red, orange, yellow, white, or blue in color.

879. Around 3.3 million lines of computer code on the ground support more than 1.8 million lines of ISS flight software code.

880. Venus is the only planet in our solar system that is named after a female.

881. Astronauts don't have much taste in space as we

rely on gravity for food to hit our taste buds. Our sense of smell also contributes to our taste. In space, fluids build up in the sinuses without gravity, another reason why food tastes bland in space.

882. Guion "Guy" Bluford Junior was the first African American in space on the space shuttle Challenger in 1983.

883. Saturn's orbit around the sun is 29.4 Earth years.

884. The second hottest planet in our solar system is Mercury. Venus is the hottest.

885. To be able to be a NASA trainee, you must be American, pass a challenging medical and physical examination, and have 20/20 vision.

886. Don Lind was selected to be an astronaut in 1966, but he didn't fly for 19 years until 1985 for various reasons such as canceled missions or he was back up or not needed.

887. All the planets together make up 0.14% of our solar system, of which 99% would be from Jupiter, Neptune, Uranus, and Saturn, the gas giants.

888. There have been 56 uncrewed missions to Mars since 1960, of which only 26 have been successful.

889. Without a spacesuit, a person can only survive for 15 to 30 seconds in space if they breathe out during this time. The lack of oxygen in space would kill them.

890. NASA's New Horizons spacecraft was launched in 2006 to fly by and study Pluto, which it did in 2015! It also managed to study another Kuiper Belt object called 2014 MU69.

891. Some astronomers would classify Pluto as a comet if it were closer to the sun.

892. The temperature through the Earth's core increases a degree every 18.2 m (60 ft).

893. Impact craters, created by the collision of space objects, exist on all the inner planets.

894. Scientists believe that the universe will end up in a Big Freeze as the world continues to cool while it expands.

895. Objects can leave and escape the Outer Event Horizon section of a black hole.

896. With a very thin and almost no atmosphere, Mercury has no weather or winds.

897. The atmospheres of Venus, Earth, and Mars can generate local weather.

898. Eris, the dwarf planet, was discovered on 21 October 2003.

899. Ancient civilizations once thought meteors were a sign of anger from the gods, or they were gifts from angels. In the 1600s, many people nicknamed them thunderstones as they thought they came from thunderstorms.

900. Earth was called the Blue Planet when astronauts first went into space and saw it covered in oceans.

901. Most astronomers believe the Big Bang caused galaxies.

902. 97% of Earth's water is salted and found in oceans. This means 3% is fresh water, but more than 2% is frozen in our glaciers, leaving only 1% fresh water in lakes, rivers, and underground.

903. 47% of Earth's crust is made up of oxygen.

904. Compared to other planets, of which it is not, Pluto has a very slow rotation and takes six days, 9 hours, and 17 minutes to rotate once. Jupiter rotates the fastest in less than 10 hours.

905. A reconnaissance satellite was designed for military intelligence.

906. There are more than 3,000 regions of star births

in the Pinwheel Galaxy's spiral arms, the most of all the spiral galaxies.

907. Giant storms occur on gas giants; for example, the Great Red Spot on Jupiter is a gigantic storm about twice the size of Earth.

908. Lightning occurs more frequently on Venus than on Earth.

909. Meteor showers happen regularly or yearly when the Earth passes through the dusty debris left by a comet or an asteroid. For example, the Eta Aquariids meteor shower that falls in May and the Orionids meteor shower that falls in October come from Halley's Comet.

910. The circumference of Neptune at its equator is 155,600 km (96,685 mi).

911. Laika was the name of the first creature (dog) to go into space in 1957. It survived a week in space on Sputnik 2 and died when the spacecraft burned up on re-entry into the Earth's atmosphere.

912. Clyde Tombaugh was 24 years old when he discovered Pluto.

913. Uranus is the only planet that spins on its side. Its axis of rotation is tilted sideways, so the north and south poles are where our equator is. It rotates once every 17 hours and 14 minutes.

914. The closest planet to the Sun is Mercury.

915. Asteroids can be anywhere from 10 m (32 ft) to hundreds of km or miles in diameter.

916. The inner planets may have no or a couple of moons. Mercury and Venus have no moons while Mars has two, and Earth has one.

917. One of Jupiter's moons, Io, has many active volcanoes on it, shooting plumes of up to 400 km

(250 mi) into its atmosphere, It is considered to be the most active moon in our solar system.

918. In 2009, a US communications satellite Iridium 33 collided with a derelict Russian satellite Kosmos 2251 in space. Both satellites were destroyed.

919. Easter is calculated as the first Sunday after the first Saturday after the first full moon after the equinox.

920. Jupiter's circumference is 439,264 km (272,946 mi), and its diameter is 11 times the size of Earth's.

921. Because Pluto is so far from the sun, it is icy. It is a rock with a very thick ice layer.

922. The Andromeda Galaxy has a double nucleus with a massive star cluster at its center and a hidden supermassive black hole at its core.

923. Mercury is the smallest planet in our solar system, and Mars is the next smallest.

924. Jupiter has the shortest day of all the planets.

925. An astronomical unit is a distance from Earth to the sun.

926. The red giant HE 1523-0901 is known as the oldest star in the solar system.

927. The densest planet and the third most massive is Neptune.

928. The Andromeda Galaxy is getting closer to the Milky way at about 100 to 140 km/s (62 to 67 m/s).

929. You need at least a 4-inch telescope to see the Pinwheel Galaxy. However, to see its amazing spiral structure, you need an 8-inch telescope.

930. A group of chain-like galaxies called Markarian's Chain to sit near the Messier 87 Galaxy.

931. The first component of The International Space

Station (ISS) sent to orbit was the Zarya module. It was launched into space on November 20, 1998, on a Russian Proton rocket. Zarya provided propulsion, control of attitude, communications, and electric power.

932. Four European ATV cargo spacecraft, four Japanese HTV cargo spacecraft, three SpaceX Dragons, 37 Space Shuttle missions, and 89 Russian spacecraft have used the ISS as a spaceport.

933. Astronauts and cosmonauts coming from 15 different nations have visited the ISS. There has been a total of 352 flights to the ISS, including 211 individuals, 31 of whom were women, and 7 were 'space tourists.'

934. Every year, a piece of asteroid (meteoroid) falls into our atmosphere, resulting in a fireball. Before reaching the ground, the meteoroid usually burns up.

935. The nearest cluster of stars in the solar system is the Beehive Cluster in Cancer constellation. The name was given as it resembled a swarm of bees.

936. Shiny dust particles left behind by comets are shooting stars.

937. John Glenn, at the age of 77 years, is the oldest person to go to space. He went as a human guinea pig so scientists could study geriatrics in space.

938. A Deimos day (one of Mars' moons) is equivalent to 2.7 Earth days.

939. Many unanswered questions stem from astronomy, such as: is there life on other planets, what is dark energy and dark matter, what is the fate of the universe?

940. Apart from Earth, Venus was the first planet to be seen from space.

941. It remains a mystery what happened before Big Bang.

942. It is not very easy to send spacecraft to Mercury due to its closeness to the Sun, so only three have been launched since 1973.

943. Isaac Newton invented the reflecting telescope during the late 1600s.

944. Many automated systems discover asteroids near Earth, but it is rare for them to cross paths with Earth.

945. Our solar system moons all have mythological characters' names except Uranus, which was named after William Shakespeare's plays and Alexander Pope's poem.

946. The second densest planet in our solar system is Mercury. It is made up of heavy metals and rocks.

947. We can only see an object in space if the light is reflected off it. That's why space is so dark.

948. An elliptical galaxy is very bright, so there would be light all day and night if Earth were situated in one.

949. A partial eclipse happens when only part of the Moon passes through Earth's shadow. It looks like a dark bite has been taken out of the Moon.

950. A reflecting telescope gives cheaper to make and provides a clearer picture than a refracting telescope. However, the optics require much maintenance.

951. A comet has two tails - a dust tail which we can see with our naked eye and a plasma tail which can be photographed but unable to be seen without equipment.

952. Strong winds blow on Saturn at speeds of over 800 kph (500 mph).

953. The surface temperature on the moon is extreme

and varies from 107 degrees C (224.6 deg F) during the day to -153 deg C (-243.4 deg F) at night.

954. Many subfields of science have come out of astrology, including planetary science, stellar astronomy, galactic and extragalactic astronomy, cosmology, and solar astronomy.

955. The top layer of Saturn's atmosphere is made mostly of ammonia ice. Underneath this layer is a layer of water ice and below these layers of cold hydrogen and sulfur ice combinations.

956. Triton's (one of Neptune's moon) atmosphere differs from that of other planets as it has a thermosphere instead of a stratosphere.

957. No person has ever seen a black hole even though we believe they exist.

958. The clouds of mercury, ferric chloride hydrocarbons, and sulphuric acid that surround Venus create the most corrosive acid rain of all planets.

959. You can combine several reflecting telescopes to make a big super telescope.

960. A red supergiant star, for example, Betelgeuse, is larger than a yellow dwarf star. A red hypergiant is even bigger than a supergiant.

961. To avoid the devastating impact of an asteroid collision with Earth, scientists have proposed that we could use nuclear explosions to break it up into smaller pieces.

962. Cygnus, the Swan constellation, is about 1,000,000 times bigger than the sun and is the biggest star in our universe.

963. Nebulae are often called because of their shape. For example, the Horsehead Nebula resembles a horse's head.

964. Neptune is the only planet that you can't see without a telescope.

965. Earth has a tilt of about 66 degrees.

966. Venus is also called the Morning and Evening Stars as it is bright at both times of the day.

967. As Pluto is about the same size as one of Neptune's moons, Triton, some scientists think it may have orbited Neptune and was pulled out of its orbit to go into Plut's.

968. The Messier 87 Galaxy has a similar size of 120,000 light-years to the Milky Way.

969. Some astronomers think that there were many big bangs in the past before the one we know about, and the only reason we're here is that we could exist it this universe!

970. The launch of the first satellite by the Soviets started the Space Race between the Soviets and the Americans.

971. The space shuttle launches vertically like a rocket and lands horizontally like an airplane.

972. After Albert 11 in 1949, Albert III, Albert IV, and Albert V were the next three monkeys to go to space, but all died either during the flight or killed on impact back to Earth.

973. You can see meteors 120 km (74.5 mi) high above Earth.

974. The Kuiper Belt is named after the astronomer who predicted its existence, Gerard Kuiper.

975. In 1978, Charon, Pluto's largest moon, was discovered by James Christy, a US Naval Observatory scientist, when he observed a very slight bulge on one side of Pluto. He suggested Charon after his wife's nickname, "Char" for Charlene.

976. In the solar system, Eris, dwarf planet, is situated past the Kuiper belt.

977. Triton (one of Neptune's moons) has the coldest temperature of all the objects in our solar system. The average temperature on Triton is -235 degrees C (-391 degrees F).

978. On 18 June 1983, the first American woman went to space in the space shuttle Challenger. Her name was Sally Ride.

979. Saturn's moon Titan has slightly higher surface pressure than Earth.

980. Venus has a similar gravitational force as Earth does just slightly less, so you would weigh less on Venus than you would on Earth.

981. Life can only exist on Earth because of the atmosphere. If there were no atmosphere, the Earth would be too hot or cold, and Earth would become similar to the moon.

982. 1991 BA, with a diameter of 6 meters (19.6 ft), is the smallest known asteroid.

983. Charon, Pluto's largest moon, has an impressive chasm, 59.5 km (37 mi) wide, named Serenity Chasm, named after the ship on Joss Whedon's cult classic show Firefly.

984. The only planet to be less dense than water is Saturn.

985. The Hubble Space Telescope can see and study Neptune.

986. Like Earth, Venus is made up of a core of iron, a rocky mantle, and a crust.

987. A fireball is a meteor that burns brighter than usual.

988. Deep inside Saturn, hydrogen becomes metallic.

989. Venus' atmosphere contains mostly carbon dioxide.

990. Scientists believe Ceres, dwarf planet, has a rocky core with an icy inner mantle 100 km (62 mi) thick. The ice mantle may contain as much as 200 cubic km of water, which is more fresh water than exists on Earth.

991. If you combine all of the asteroids in the Asteroid Belt, it would be about the size of our moon.

992. Tides and weather on Earth are predictable when the moon is farthest from Earth. When the moon is closer, its greater gravitational pull creates bigger waves and more unstable weather.

993. Luna 9 was the first spacecraft that landed on the moon and helped astronomers to understand that the moon had a stable surface for landing.

994. Between 6 Apollo crews, 385 kg (850 lb) of the moon was brought back to study.

995. Uranus has 27 moons, all named after William Shakespeare and Alexander Pope's characters.

996. The thin, cold atmosphere becomes thicker and hotter approaching Jupiter's core, gradually turning into a thick, dark fog. About 1000 km (621 mi) down in the blackness, pressure squeezes the atmosphere so hard that it becomes liquid.

997. The life cycle of a star begins in a cloud of dust called a nebula.

998. Saturn's moons are all frozen.

999. Titan is Saturn's largest moon, out of its 62 moons.

1000. The Apollo 14 commander, Alan Shepherd, hid a golf club on the spacecraft, and he was the first person to play golf on the moon.

1001. The sun has a radius of about 695,508 km (432,168 mi), which is about 109.2 times bigger than Earth's.

1002. A yellow dwarf star, such as the Sun, will

eventually run out of hydrogen fuel and become a red giant.

1003. Uranus' inner moons consist of half water ice and half rock. Scientists don't know the composition of the outer moons, but they believe them to be captured asteroids.

1004. An asteroid about the size of a car hits Earth's atmosphere every year but burns up before it hits land.

1005. The Cancer Constellation used to be named 'the crayfish' 3,000 years ago in Babylonian times.

1006. The ISS Cupola module has a 7-window observatory area that was compared to the Millennium Falcon's 'turret' in the Star Wars movie.

1007. NASA admitted in 2006 that they had taped over the original tapes of the moon landing hence speculations that the arrival was not real.

1008. Refractor lenses are also used in binoculars and gun scopes.

1009. Astronauts live inside pressurized modules on The International Space Station (ISS).

1010. Scientists debate whether Charon was formed when it collided with Pluto or from a collision between Pluto and a Kuiper object (like how Earth and its moon were created).

1011. Asteroids are blown-out comets! When the ice melts, only the solid material remains.

1012. The coldest world in the solar system is Triton, one of Neptune's 14 moons.

1013. For an artificial satellite to keep its orbit and stay in space, it must travel faster than 28,200 kph (17,500 mph).

1014. The Triangulum Galaxy is considered as an

isolated galaxy because it has not interacted with other galaxies recently.

1015. Gas giants are not all gas; for example, Jupiter and Saturn both have layers of molecular and liquid metallic hydrogen.

1016. A reflector is another name for a reflecting telescope.

1017. Halley's Comet is now a Kuiper Belt object but thought to have come originally from the Oort Cloud.

1018. A dwarf planet that orbits outside of Neptune is known as a plutoid. Pluto, Haumea, Eris, and Makemake are all plutoids. Ceres is not.

1019. An elliptical galaxy is a group of stars that bunch together to form an oval shape, a stretched out and elongated circle.

1020. Because Earth is flattened at the poles and bulges at the Equator, it is not round. It is the geoid, an elliptical spheroid shape.

1021. Gamma-ray astronomy is the use of the shortest wavelengths to study space objects.

1022. Fourteen astronauts were killed in the Columbia and Challenger space shuttles.

1023. The Tropic of Cancer is named after the Cancer Constellation. It is the most northern latitude reached by the sun.

1024. Fifty-two computers control ISS.

1025. The space station is nearly four times larger than Mir, the Russian space station, and about five times larger than the US Skylab.

1026. The Hubble Space Telescope was given its name after Dr. Hubble, famous for providing evidence of the Big Bang Theory.

1027. A blue moon is the second full moon in a calendar month. It comes from an old English word 'belewe' or 'betrayer' as people viewed a full moon before lent was a 'betrayer moon.' 'Belewe' eventually became 'blue.'

1028. The moon's gravity is about 1/5 of Earth's.

1029. There is no sound in space.

1030. The largest galaxy that we know of in the universe contains over 100 trillion stars, compared with the Milky Way that is estimated to have about 400 billion stars. We can only see about 2,500 of these stars from Earth.

1031. Almost identical solar eclipses happen every 18 years and 11 days. This period is called a saros cycle.

1032. Nuclear fusion, when hydrogen burns to make helium, occurs in a star's core and creates a star's energy.

1033. The Hubble Space Telescope has mirrors to capture images, so it is known as a reflector telescope.

1034. To travel from Earth to Charon would take 4.6 light-years. This would equate to 6,293 years if you were driving a car at 104 kph (65 mph) or 680 years if you're flying a Boeing 777 going 949 kph (590 mph). It took New Horizons 9 years going at 80,467 kph (50,000 mph), which is the fastest speed for a launched spacecraft.

1035. An artificial satellite has two important parts - the antenna to send and receive information and a power source, like a battery or a solar panel.

1036. The Hubble and Spitzer Space telescopes have observed that most of the star formation in the Sombrero Galaxy occurs at the outer tip of the dust ring.

1037. Just like Pluto, Charon takes 248 years to orbit the sun.

1038. A killer satellite was designed to destroy warheads.

1039. Vesta, larger than 500 km (310 mi) in diameter, is the only asteroid that you can see with the naked eye.

1040. Haumea has an equatorial diameter (from one side to the other side passing through the center) of 1,960 to 1,518 km (1,217 to 943 mi). It has a polar diameter of 996 km (618 miles).

1041. Collisions and impacts from the material may have impacted the orbits of Titan and Saturn's other moons intro their current positions.

1042. The ISS is arguably the single item most expensive ever built. As of 2010, the station's cost is estimated to be $150 billion.

1043. The stars that orbit the Milky Way travel at very fast speeds, much faster than Earth's orbit around the Sun. If Earth traveled at similar speeds to these stars, we would orbit the Sun in 3 days, not 365!

1044. Occasionally scientists have observed strange colored lights on the moon, which are thought to be escaping gases from inside the moon.

1045. The Russians sent the first space mission to Venus. Venera 3 successfully landed on Venus in 1966 but crashed and could not send any data back.

1046. Our moon is tilted 20 - 30 degrees as it orbits Earth. The moons from other planets orbit the equator of the planet.

1047. As a star is getting to the end of its life, it changes color, density mass, and size.

1048. The ancient Babylonians were the first ones to record their Jupiter sightings.

1049. 23% dark matter, 4% ordinary matter, and 73% dark energy make up our universe.

1050. The Cancer constellation, a medium-size

constellation with an area of 506 square degrees, is the 31st largest.

1051. Our Universe contains about 70 thousand million, million, million stars.

1052. The universe started as a small hot ball that cooled as it expanded, according to the Big Bang theory.

1053. The Andromeda Galaxy is also known as the Great Andromeda Nebular and Messier 31 or M31.

1054. A meteoroid in space becomes a meteor when it enters Earth's atmosphere, which then becomes a meteorite when it hits Earth's surface.

1055. Scientists believe that Haumea collided with a large object billions of years ago, resulting in two moons being formed.

1056. The moon is 384 403 kilometers (238 857 miles) from Earth.

1057. The temperature of Earth's core is about 4,300 degrees C (7,772 degrees F).

1058. While only seven astronauts can fit on a space shuttle at any one time, it has flown over 600 astronauts into space.

1059. Earth takes 23 hours, 56 mins, and 4 seconds to rotate around its axis (not 24 hours), i.e., one day is 4 minutes shorter than we think.

1060. Fruit flies that were sent along with corn seeds were the first living things sent to space in 1947.

1061. Pluto's moon, Charon has freezing temperatures of -220 degrees C (-364 degrees F) compared with the coldest object in the solar system, Neptune's moon, Triton, with a temperature of -235 degrees C (-391 degrees F)

1062. Earth is made of 32.1% iron, 30.1% oxygen, 15.1% silicon, and 13.9% magnesium.

1063. Europa, Jupiter's moon, is about 780 million km (485 million mi) from the Sun.

1064. The size of Kuiper Belt Objects (KPO) is challenging to measure as they are so far away. Their reflectiveness determines them - the infrared observations are measured by the Spitzer Space telescope, which gives an estimate of the size.

1065. Astronaut means 'star sailor.'

1066. Supernovas can be created by nuclear reactions or when stars collapse after running out of fuel.

1067. Mark, which comes from the Latin word for Mars, was the name of the character that Matt Damon played in the 2015 movie, The Martian.

1068. Messier 87's mass is 6,600,000,000 times bigger than the suns.

1069. The Roman and Greek ancient civilizations thought Venus was two different objects. As Venus is closer to the Sun and has a shorter orbit, it overtakes Earth's orbit and can be seen at sunset and sunrise. The Greeks called the two objects Phosphorus and Hesperus, and the Romans gave them the names Lucifer and Vesper.

1070. A spiral galaxy eventually burns through its gases. As their dust start formation slows down, they lose their spiral shape and become an elliptical galaxy.

1071. A penumbral lunar eclipse happens when part of the outer surface of Earth's shadow falls on the Moon. It is harder to observe a penumbral lunar eclipse than a partial or total eclipse.

1072. In 1781 Sir William Herschel, a British musician and sky watcher discovered Uranus.

1073. The only planet that we know of that has free oxygen, liquid water, and life is Earth.

1074. Some people think that the moon gives off light, but it doesn't - it just reflects the sun's light.

1075. Most meteorites have been found in Antarctica than anywhere else on Earth.

1076. A star explodes in the universe every second.

1077. The Sun comprises three-quarters hydrogen and helium.

1078. The official name for Earth's moon is the Moon (with a capital M).

1079. The circumference of Mercury is 15,329 km (9,525 mi).

1080. It takes Uranus 84 Earth years to orbit the Sun.

1081. Uncrewed spacecraft have reached every planet or its orbit.

1082. When a supernova occurs, it shoots billions of atoms in every direction, which form nebulae made up of clouds of dust, gases, hydrogen, and helium.

1083. Copernicus suggested that the Sun was the center of the universe after Europa and three other Galilean moons, Lo, Ganymede, and Callisto, were discovered.

1084. AU stands for Astronomical Unit. One AU is the distance between the sun and Earth.

1085. One of Deimos' largest craters with a diameter of 1,609 km (1,000 mi) is named Swift. It was named after the author of the book Gulliver's Travels, Jonathan Swift. Jonathan wrote about Mars' two moons 151 years before they were discovered!

1086. Ganymede is the third of the Galilean satellites. It orbits Jupiter every seven days and 3 hours, at a distance of 1,070,400 km (665,000 mi) from the planet.

1087. Some bacterias, such as salmonella and E-coli, grow faster in space.

1088. As of 2018, approximately 8,378 satellites have been launched into space since the first one in 1957. Less than 40% are operational, i.e., only 1,957 of the 8,378.

1089. Eclipse means 'downfall,'

1090. Eris' moon, Dysnomia, is named after Eris' daughter, the demon goddess of lawlessness, in Greek mythology.

1091. Twice daily, Phobos, one of Mars' moons, rises in the west and sets in the east.

1092. The Outer Space Treaty rules that the moon can be used for peaceful purposes by everyone, and it prohibits the use of weapons of mass destruction or military bases to be built on the moon. So there will never be a war on the moon!

1093. Mars' gravity is 37.5% less than Earth's, so if you weighed 45 kg (100 lb) on Earth, you would only weigh 17 kg (38 lb) on Mars! You could also jump three times higher!

1094. Because there are more oceans than land on Earth, most meteors fall in the sea.

1095. Ancient civilizations didn't discover Uranus as it was not bright enough (and they hadn't invented telescopes yet).

1096. Ceres is considered not to have an atmosphere. However, a Herschel space telescope in 2014 showed that it might have water vapor in its atmosphere. This could be explained by sporadic ice being ejected by small impacts.

1097. Charon, Pluto's largest moon, has grown bigger in the past as its surface stretched.

1098. Many stories have been told about 'the man in the moon' who was reportedly placed there for stealing. What he took has been a topic of debate for years!

1099. A comet can break up if it gets close to the Sun too many times or if it comes too close to the Sun or another planet in its orbit.

1100. A comet is made up of frozen water, supercold methane, ammonia, and carbon dioxide ices mixed with rock and dust and other debris from the solar system.

1101. The Milky Way was also described in Greek mythology as the road to Mt Olympus.

1102. A person who is trained by a spaceflight program to lead, pilot, or be a crew member on a space expedition is called an astronaut or a cosmonaut.

1103. It is safe to watch a lunar eclipse as the Moon isn't giving off its light.

1104. The electric power system on the ISS is connected by more than eight miles of wire.

1105. Objects in the Oort Cloud travel in all different directions and don't follow a typical orbit around the sun. That's why it's called a 'cloud' and not a 'belt' like the Kuiper Belt.

1106. There are 12 zodiac constellations: Aries, Taurus, Gemini, Cancer, Leo, Virgo, Libra, Scorpius, Sagittarius, Capricornus, Aquarius, and Pisces.

1107. With the naked eye, the ISS can be seen from all around the world, appearing as a slow-moving, bright white dot in the night sky.

1108. Callisto, Jupiter's second-largest moon, has an interior ocean made of water and ammonia. Its surface is made up of water and carbon dioxide ice, rocks, and silicate dust and hydrocarbon compounds.

1109. In 1996 the Hubble Space Telescope found a thin layer of oxygen on Ganymede, one of Jupiter's moons. The thin atmosphere is not enough to sustain life.

1110. The climate on Neptune is very active, with massive storms in its upper atmosphere and strong winds blowing up to 600 meters per second (1,968 miles per sec).

1111. The shape of the Pinwheel Galaxy is distorted from its interactions with its satellite galaxies.

1112. The methane in Neptune's atmosphere absorbs all the red light, so Neptune appears blue.

1113. Neptune's thin rings comprise ice particles and dust, which are possibly covered with a carbon type substance.

1114. Astronauts must exercise at least 2 hours every day in space to maintain their muscle mass.

1115. The first artificial satellite in the world was Sputnik 1 from the Soviet Union.

1116. There are 19 constellations in the Hercules family, 10 in the Ursa Major Family, 9 in the Perseus Family, and 5 in the Orion Family.

1117. The length of Makemake's day, about 22.5 hours, is similar to Earth's.

1118. The average mission on a satellite is 3 to 4 years. They are sometimes re-orbited after this time, but most of them are sent to the graveyard orbit.

1119. Uranus is the coldest planet with an average temperature of -197 to -224 degrees C (-322 to -371 degrees F).

1120. Uranus receives about 1/400th of the Sun that Earth gets.

1121. Jupiter's upper atmosphere is divided up into cloud belts and zones. They are primarily made of ammonia crystals, sulfur, and mixtures of the two compounds. Below the massive atmosphere, there are layers of compressed hydrogen gas, liquid metallic hydrogen, and a core of ice, rock, and metals.

1122. Saturn is a gas giant with a radius of about nine times longer than Earth.

1123. Eight spacecraft have visited Jupiter, including Pioneer 10 and 11, Voyager 1 and 2, Galileo, Cassini, Ulysses, and New Horizons missions.

1124. The inventor of the Super Soaker water squirt gun was Lonnie Johnson, a NASA scientist.

1125. The moon orbits Earth in an anticlockwise direction.

1126. An Einstein-Rosen Bridge is better known as a wormhole. This is a hypothetical tunnel that would join two different points in space, perhaps two different universes.

1127. The second-largest planet in our solar system is Saturn. The largest is Jupiter.

1128. All the 12 people who have walked on the moon have said moon dust smells like gunpowder.

1129. Venus has mountains, valleys, and lots of volcanoes. It has lots of larger craters as smaller asteroids don't make it through the atmosphere.

1130. Seven of the eight planets are named after Roman gods or goddesses. This convention had continued since ancient times when Mercury, Venus, Mars, Jupiter, and Saturn were discovered.

1131. The descent module of Mars 2, a space probe launched by the Soviet Union in 1971, separated from the orbiter module about 4.5 hours before it reached Mars. The descent module malfunctioned and crashed, becoming the first human-made object to touch Mars.

1132. The closest star in the Milky Way to Earth is called the Proxima Centauri. It is over four light-years away (1 light year = 10 trillion kilometers or 5.9 trillion miles).

1133. The universe is approximately 13.7 billion years old.

1134. The Cosmic inflation theory attempts to explain the exponential expansion of space in the early universe.

1135. The Noctis Labyrinthus, a region on Mars containing the largest labyrinth of valleys and canyons, means 'labyrinth of the night.'

1136. Triton's (one of Neptune's moon) surface looks like a rockmelon. Its cantaloupe terrain covers most of the western half of the moon with craters that are all about the same size so unlikely to be impact craters.

1137. Earth travels through space at 107,826 kph (67,000 mph), so every hour, you are 107,826 km (67,000 mi) farther away!

1138. The HII regions of the Pinwheel Galaxy can create hot superbubbles, hundreds of light-years wide, from the high number of bright hot young stars in them.

1139. The Milky Way has a long bar in the center from where the spiral arms spin out.

1140. Saturn is named after the Roman god Saturnus.

1141. Eris is three times further away from the sun than Pluto at a distance of 96.4 astronomical units (more than 14 billion km). It takes sunlight more than 9 hours to travel to Eris.

1142. Scientists believe that the universe that we can observe is about 13.8 billion years old.

1143. The average distance between stars in the Milky Way is about five light-years or 48 trillion km (30 trillion miles).

1144. The inner planets are known as rocky planets, while the outer ones are known as gas giants.

1145. Uranus is pronounced 'YOU-ra-nus' not 'YOUR-anus.'

1146. The only spacecraft to fly by Neptune is the Voyager 2 in 1989.

1147. Charon, Pluto's largest moon, is not oblong shaped like Hydra and Nix. Due to its gravity, Charon collapsed into a spheroid shape.

1148. The thin atmosphere of Neptune's moon, Triton, has tiny amounts of carbon monoxide and methane.

1149. A satellite moves very fast, at speeds of about 28,968 kph (18,000 mph) so they can orbit Earth 14 times a day.

1150. It is believed that Galileo's blindness is due to looking directly into the sun with his telescope. But the truth is he became blind from cataracts and glaucoma.

1151. We wouldn't be able to see the moon and its phases if there wasn't a sun.

1152. Charon, Pluto's largest moon, has no atmosphere.

1153. NASA's Hubble Space Telescope is the first astronomical observatory placed into orbit around Earth.

1154. Earth could fit inside the Sun 1.3 million times due to the Sun's immense size.

1155. In 2006 the Hubble Space Telescope released the largest and most detailed image of the Pinwheel Galaxy.

1156. While there is no water on Mercury's surface, there may be underground water.

1157. The first supernova that was observed and recorded was by Chinese astronomers in 185 AD.

1158. Many asteroids and comets have collided with the moon in the past, creating many impact craters.

As the moon has no atmosphere, it has no weather to erode these craters.

1159. On 14 June 1949, Albert 11 became the first monkey to go to space.

1160. Astronomy is useful to determine seasons to assist farmers with when to plant and harvest their crops.

1161. Earth has one moon and two more asteroids that also orbit Earth. Earth's two co-orbital satellites are called 3753 Cruithne and 2002 AA29.

1162. The space shuttle was 56 m (184 ft) long and weighed 2 million kg (4.5 million lb).

1163. Saturn's thin rings, about 20 m (65.5 ft) thick, extend more than 282,000 km (175,000 mi) from the planet.

1164. Although the Magellanic Clouds are irregular galaxies, they are often referred to as spiral galaxies as they have a bar in the center of the galaxy.

1165. About 500 active satellites are considered Low Earth Orbit satellites as they orbit Earth under 2,000 km (1,240 mi) away.

1166. Lunatic comes from the word 'lunar,' and the phases of the moon have been associated with madness. Aristotle believed that the full moon affected the water in a person's brain and made them insane!

1167. Only three planets spin anticlockwise - Pluto, Venus, and Uranus.

1168. Every time a large comet orbits the Sun, it loses about 1 to 3 meters (3 to 10 feet) off the surface of its nucleus. Halley's Comet will eventually lose its tail once a day and may disappear!

1169. The surface of Mars has extreme radiation every time the Sun rises as it has no ozone layer.

1170. The gravitational pull of the moon creates our tides.

1171. Neptune is known as a small gas giant - it contains mostly gas.

1172. As there is no magnetic field on the moon, a compass wouldn't work.

1173. Deimos' orbit is moving away from Mars while Phobos is getting closer and will collide with Mars in about 50 to 100 million years. In the future, Deimos will move out of Mars' gravity and will no longer orbit the planet.

1174. The polar ice caps on Mars are made up of frozen carbon dioxide (dry ice).

1175. Galaxy means 'milky.'

1176. From Earth, the moon and the sun look the same size as the moon is 400 times smaller than the sun and also 400 times closer to Earth.

1177. The Apollo 11 astronauts couldn't get insured from their trip, so they signed photos of themselves and left them for their families to auction - in case they didn't return.

1178. Mars was named after the Roman God of war due to its reddish color resembling blood.

1179. The surface of Europa, Jupiter's moon, is about 20 to 280 million years old, even though it is about 4.5 billion years old.

1180. It's easier to see meteors at night.

1181. Saturn has the most moons in the solar system. Some are awaiting confirmation. Jupiter is next.

1182. A very heavy star that has exploded into a supernova can become a black hole.

1183. In 1612 Galileo Galilee drew Neptune, but the drawings were of a fixed star and not the planet, so he is not said to have discovered Neptune.

1184. Deimos (Mars moon) has an irregular shape and not round like the other moons.

1185. The volume of the Earth's moon is equivalent to the size of the Pacific Ocean on Earth.

1186. When a red dwarf star or a red giant starts to die, it will turn into a small white dwarf star until they stop emitting white light and dies as a black dwarf star.

1187. Deimos (Mars moon) is named after the Greek god of war, who was a son of Ares (Mars) and Aphrodite (Venus).

1188. Earth's rotation is slowing down gradually, decelerating at about 17 milliseconds per hundred years. This means our days are getting longer. It could take up to 140 million years, however, before our 24 hour day becomes 25 hours.

1189. The two Magellanic Clouds, known as the Large and Small Magellanic Clouds, are irregular dwarf galaxies that can be seen in the Southern Hemisphere.

1190. Jupiter is named after the king of the Roman gods. To the Greeks, it represented Zeus, the thunder god. The Mesopotamians saw Jupiter as the god Marduk and patron of Babylon's city. Germanic tribes considered this planet to be Donar or Thor.

1191. The five space shuttles were named Enterprise, Columbia, Challenger, Discovery, and Atlantis.

1192. Mercury's orbit around the Sun ranges from 46 to 70 million km (28.5 to 43.5 million mi) from the Sun.

1193. Saturn has ammonia crystals in its upper atmosphere, so it looks pale yellow.

1194. A dark nebula, for example, the Horsehead

Nebula, blocks all the light from interstellar grains of dust so you can't see anything behind it.

1195. The Whirlpool Galaxy is about 43% of the size of the Milky Way.

1196. The ancient Assyrians called Saturn 'Lubadsagush', meaning 'oldest of the old' as it was so so slow.

1197. The term for an orbit around Earth is 'geocentric.' An orbit around the sun is referred to as 'heliocentric,' and one orbiting Mars is called 'Areocentric' (after Ares, which is another name for Mars).

1198. A person would weigh the lightest on Pluto and heaviest on Jupiter.

1199. The cosmic microwave background radiation's temperature is 2.7 degrees Kelvin (-270.45° C, -454.81° F).

1200. Uranus rotates in a backward direction, similar to Venus but different to all the other planets, including Earth.

1201. When Eris and Pluto were both demoted to dwarf planet status, there remained only eight planets in our solar system.

1202. Charon remains permanently in one place in Pluto's sky ('tidal locking'), never rising or setting as it takes to the same time to orbit Pluto as Pluto does to orbit the sun. The same surfaces always face each other.

1203. Uranus has also known as the 'most boring planet in our solar system.'

1204. An object in the Kuiper Belt or the Oort Cloud is called a Trans-Neptunian object.

1205. NASA is working on developing a nuclear-propelled rocket to go to space, which would mean that it would reach Mars in half the amount of time.

1206. Uranus is the seventh planet from the Sun.

1207. The sun is white, not yellow. The Earth's atmosphere makes it look yellow.

1208. Neil Armstrong, the first man to step on the moon, was late in submitting his NASA application by a week. His friend had to slip the form into the pile so that it was accepted.

1209. Uranus is known as one of the ice giants.

1210. The fifth mass extinction on Earth could be due to the Ordovician-Silurian extinction events resulting from a supernova between 447 and 443 million years ago.

1211. The Spitzer Space Telescope can no longer take pictures under extreme temperatures as it has run out of liquid helium.

1212. In the Local Cluster, Andromeda is the largest galaxy, but it doesn't have the biggest mass. The Milky Way is more massive as it contains more dark matter than the Andromeda Galaxy.

1213. Ceres is the closest dwarf planet to Earth.

1214. In the 1100s, Ptolemy, a Greek astronomer, recorded The Cancer constellation.

1215. Sunlight takes 20 times longer to reach Uranus than Earth.

1216. A delta rocket launched the Spitzer Space Telescope, and it weighed around 929 kg (2,049 lb).

1217. Vader Crater is a crater on Charon named after Darth Vader from Star Wars.

1218. Jupiter, Saturn, Uranus, and Neptune are the four gas giants in our solar system. You would not be able to walk on them as they are made up mostly of gas.

1219. The Milky Way contains up to 400 billion stars, maybe just as many planets and has a diameter of about 120,000 light-years.

1220. The darkness during a total lunar eclipse is measured using the Danjon Scale. It ranges from 0 when the Moon is almost invisible to 4 when the Moon is a very bright yellowish-orange color.

1221. A crater, 9.5 km (6 mi) wide, covers most of Phobos (one of Mars' moons). It was named Stickney, after the wife of the person who discovered it, Chloe Angeline Stickney.

1222. A meteor shower is often caused by debris coming off a broken comet.

1223. NASA's Pathfinder, launched in 1996, made the first airbag mediated touchdown on Mars.

1224. A total solar eclipse is rare and occurs about once every 1 to 2 years. In contrast, partial solar eclipses can happen 2 to 5 times every year.

1225. The Andromeda Galaxy belongs to the Local Group cluster of galaxies.

1226. In the middle of the Triangulum Galaxy is a nebula made up of a cloud of gas and dust where stars are formed.

1227. 10 asteroids were discovered by 1849, 100 by 1868, 1,000 by 1921, 10,000 by 1989, and 700,000 by 2015.

1228. Scientists believe Olympus Mars, a volcano on Mars that is billions of years old, is still active.

1229. Charon has more craters in its northern hemisphere than in its southern hemisphere. This means the north hemisphere is older than the southern hemisphere, or the south side has undergone some resurfacing that may have buried old craters.

1230. If you weighed 100 kg (220 lb) on Earth, you would weigh 38 kg (83.6 lb).

1231. A dwarf planet is considered by the International Astronomical Union to be an object in our solar system that is not as large as a planet but is bigger than a small object in the solar system such as a comet or asteroid.

1232. The astrological symbol for Charon is a floating circle on top of a crescent, resembling the name's mythical meaning of a boatman going across the River Styx.

1233. The refracting telescope is more natural to observe the planets and the moon.

1234. Ganymede's (one of Jupiter's moons) surface is 40% dark craters, which scientists believe were caused by the substantial impact of comets and asteroids about 4 billion years ago.

1235. There are several theories about how Mars' 2 moons, Phobos, and Deimos, were created. Some astronomers have suggested that Jupiter's gravity pushed them into orbit around Mars. Others think that they were formed when gravity pushed rock and dust together. Another theory is that Mars may have collided with an existing moon, and the debris formed Phobos and Deimos.

1236. Ganymede, one of Jupiter's moons, is the largest moon in our solar system and the only moon with a magnetosphere, i.e., it has a strong magnetic field. Scientists believe its nickel and iron core generates the magnetic field.

1237. There are three main parts to a space shuttle - the orbiter, the external fuel tank, and the solid rocket boosters.

1238. Lightning strikes our planet more than 8.6 million

times a day. Each bolt of lightning contains about 1 billion volts of electricity.

1239. In ancient times, philosophers used to think that the gods were angry at them and made the Moon disappear during an eclipse.

1240. In the video game Destiny, Ceres was colonized by an alien race called the Fallen and destroyed by a civilization of post-humans who inhabit the Asteroid Belt.

1241. After observing comets to be 'stars with hair,' the Greek philosopher Aristotle gave the name comet, which means 'hair of the head' in Greek.

1242. According to scientists, the universe is expanding out at exponential rates.

1243. From 1994 to 2011, 3 supernovae have been observed in the Whirlpool Galaxy.

1244. The Milky Way is called Akash Ganga in Sanskrit, meaning 'Ganges of the heavens.'

1245. Earth has a molten metal core, which results in a magnetic field; Mars doesn't have one, so it doesn't have a magnetic field today, but evidence exists that it may once have a magnetic field. Not only did it have one, but it also reversed just like how Earth's magnetic field changes every few thousand years.

1246. Despite being one of Pluto's moons, Charon doesn't orbit Pluto. They both orbit a common center of gravity called a barycentre. Due to its large size and this orbit fact, some astronomers think that Charon should be considered a dwarf planet.

1247. The mass of Venus is 4,867,320,000,000,000 billion kg, 0.815 times the mass of Earth.

1248. The Venus Express was the first European spacecraft to orbit Venus. It was launched on 9

November 2005 and officially ended on 14 December 2014. The last signal received from the Venus Express was on 19 January 2015.

1249. The last landing of the space shuttle was on 21 July 2011. NASA wanted to focus on cheaper methods to explore space, so retired the space shuttle program.

1250. In ancient times, Venus had different names. The Egyptians referred to it as Bonou, which means 'bird,' to the Chinese it was the Tai-pe or 'the beautiful white one' and to the Chaldeans, 'the bright torch of heaven.'

1251. An extra-galactic nebula is one that exists beyond the Milky Way.

1252. The thickness of the central bulge of the Milky Way is 10,000 light-years.

1253. Jupiter has four moons - Europa, Io, Ganymede, and Callisto.

1254. Pluto is named after the Roman god of the underworld. His father was Saturn, and his brothers Jupiter was the god of the sky, and Neptune was the god of the sea.

1255. The ISS supports over 100,000 people working in 16 countries and 37 states across the US.

1256. Mars is 54.7 million km (34 million miles) away from Earth and is our closest neighbor after our moon and Venus.

1257. Dark matter or dark energy are the parts of the universe that scientists can't see or detect.

1258. Persephone, Pluto's wife, was one of the names considered for Eris, the dwarf planet.

1259. Eris' (dwarf planet) temperature is about -217 to -242 degrees C (-359 to -405 degrees F).

1260. Eris, dwarf planet, was nicknamed Xena for a short period, after the television warrior.

1261. The inner Oort Cloud starts about 2,000 Astronomical Units from the sun.

1262. Our Moon is the largest satellite of all planets in the solar system.

1263. The moon contains small amounts of water.

1264. Astronauts from the USA need to learn Russian so they can run the ISS in Russian.

1265. No spacecraft has landed on Europa, although many have visited it. The Galileo spacecraft that was launched in 1989 traveled to Jupiter and its moons.

1266. The IAU stipulates that Kuiper belt objects must be named after mythological beings. Haumea is named after the Hawaiian goddess of fertility.

1267. A spiral galaxy is categorized by how tight their spiral arms are.

1268. Although Europa is only one-fourth the diameter of Earth, its subsurface ocean may contain twice as much water as all of our oceans combined. This makes it a widely considered place to look for life outside of Earth.

1269. In 1951 a monkey named Yorik and 11 mice became known as the first animals to survive a trip to space.

1270. The Outer Event Horizon, the Inner Event Horizon, and the Singularity are the three main parts of a black hole.

1271. Comet orbits tend to be elliptical.

1272. Many scientists believe that dinosaurs became extinct about 65 million years ago when an asteroid hit Earth.

1273. Astronomy means 'law of the stars' in Greek.

1274. Pluto's atmosphere is made up of nitrogen with a little bit of carbon monoxide and methane.

1275. Astronauts stay on the ISS for about six months before returning to Earth.

1276. Astronomers study the Whirlpool Galaxy to understand the structure of galaxies and their interactions.

1277. NASA formed the Planetary Defense Coordination Office to monitor hazardous objects that come within 8 million km (5 million mi) of Earth.

1278. Neptune's surface temperature is freezing at -201 deg C (-329.8 deg F).

1279. Before Aristotle, who lived from 384 to 322 BC, the world thought the Earth was flat. Aristotle believed the world was round, and everything revolved around Earth in outer space.

1280. The diameter of our moon is 3,475 km (2,159 mi), four times smaller than Earth's diameter.

1281. About 2,600 BC, Stonehenge was built and may have been the first space observatory every.

1282. A large star won't last long in a spiral galaxy as it continually burns enormous amounts of fuel.

1283. Peggy Whitson set a record on September 2, 2017, 665 days spent aboard the ISS.

1284. Halley's Comet's nucleus is small and measures about 15 km (9.3 miles) long, 8 km (5 miles) wide, and 8 km (5 miles) thick. In comparison, the coma can stretch up to 100,000 km (62,137 mi).

1285. The space shuttle Discovery grew roses, and their scent was then used for the smell of a perfume "Zen."

1286. The distance that light can travel in one year is called a light-year.

1287. In the 3rd century BC, Aristarchus of Samos became the first person to estimate the distance and size of the moon and the sun. He was also the first person to create an ancient tool called an astrolabe, which was used to solve problems relating to time and the position of the sun and stars.

1288. The moon is moving further away from Earth - 4.6 years ago when it was formed, it was only 22,530 km (14,000 miles) away, but now it is 450,000 km (280,000 mi) away.

1289. Saturn has a smaller magnetic field than Jupiter, but its 578 times more powerful than Earth's. The big magnetic fields mean Saturn has high levels of radiation.

1290. A lunar eclipse is easier to see than a solar eclipse.

1291. The Sombrero Galaxy is smaller than the Milky Way and has a diameter of 49,000 light-years.

1292. The Spitzer Space Telescope was predicted to survive around 2.5 years in space, but it lasted for more than 11 years!

1293. Neither Venus or Mercury has an orbiting moon. All other planets do.

1294. The black holes located in the center of a galaxy is about a billion times heavier than the sun.

1295. Jupiter is the fastest rotating planet in our solar system. Its days are only ten hours long.

1296. A spiral galaxy is also called a disk galaxy.

1297. The Spitzer Space Telescope was around 720 million US dollars!

1298. Based on the size of our moon, 49 of them could fit into Earth.

1299. The Milky Way is a group of stars, gas, dust and other matter, about 100,000 to 120,000 light-years in diameter, brought and kept together by their mutual gravitational pull.

1300. You're looking back in time when you look at the

stars as their light takes millions of years to reach Earth.

1301. Asteroids used to be common in the past, but not so much anymore.

1302. The furthest an astronaut has gone into space is 401,056 km (249,205 miles) from Earth. It was on the Apollo 13 that Jim Lovell, Jack Swigert, and Fred Haise made this trip.

1303. In 2006 the Venus Express spacecraft found more than 1,000 large volcanoes on Venus.

1304. Russia plans to send humans to Mars from 2040 to 2045.

1305. The brightest star in the solar system, R136a1, located in the Large Magellanic Cloud, shines 8.7 times brighter than the sun.

1306. The Triangulum has 40 billion stars, which is a small number when compared to the Milky Way's 400 billion stars.

1307. Charles Messier first found the Whirlpool Galaxy in 1773.

1308. The Perseids is a large meteor shower that can be seen in the northern hemisphere about August every year. The Perseids contains rocks from the Swift Tuttle comet.

1309. The wrinkles that are seen on the surface of Mercury, ranging up to 1.6 km (1 mi) high and hundreds of kilometers or miles long are called Lobate Scarps.

1310. Venus is so bright it can be seen during the day with a clear sky.

1311. A comet is a small solar system object that orbits the Sun.

1312. The Cancer Constellation has the following main stars: Al Tarif, Acubens, Asellus Australis, Asellus Borealis, and Iota Cancri.

1313. Scientists think that the Milky Way is absorbing a smaller galaxy known as the Sagittarius Dwarf Galaxy.
1314. Earth only has one natural satellite, the Moon.
1315. The bright moon reflection you can see on a body of water is called moon glade.
1316. The ancient Greek philosophers were the first people to develop a model of the universe.
1317. Time does not exist in a black hole.
1318. An observatory is a remote location that has at least one large telescope.
1319. Earth's Moon stabilizes our climate, creating seasons. Without the Moon, we would not be able to live on Earth.
1320. The fifth brightest object in our solar system is Saturn.
1321. NASA's Mariner 4 was the first spacecraft to successfully fly by Mars in 1965, taking 228 days to reach the planet. The images it sent back didn't show any oceans or vegetations which scientists had hoped to find. In 2008 scientists obtained evidence that suggested liquid water and maybe life once existed on Mars.
1322. Mercury's surface has wrinkles from the planet cooling and contracting over time.
1323. A star is called a brown dwarf star if it doesn't get hot enough to cause a nuclear fusion at its core. It's not a proper star.
1324. Uranus' seasons can go for 20 years.
1325. Four of Jupiter's moons are larger than Pluto!
1326. You can see the Andromeda Galaxy as part of the Andromeda Constellation in the northern sky.
1327. The first vehicle to travel on the moon is the Soviet robot Lunokhod 1 pm 17 November 1970.
1328. The eight moon phases are New Moon, Waxing

Crescent, Crescent, First Quarter, Waxing Gibbous, Full Moon, Waning Gibbous, and Last Quarter.

1329. Solar eclipses happen more frequently than lunar eclipses.

1330. In the 16th century, Nicolaus Copernicus presented the first heliocentric model of the universe, which explained how the sun was the center of the universe and planets revolved around it.

1331. There is speculation among scientists that there may be a Martian 'Bermuda triangle' or a 'Great Galactic Ghoul' that has eaten about two-thirds of the spacecraft that we have sent to Mars.

1332. As of 2013, the construction and maintenance of the ISS were supported by 174 spacewalks outside of the modules, which is nearly 1,100 hours (46 days).

1333. Ceres is the smallest dwarf planet and the first to be discovered and visited by a spacecraft. It is found in the other Solar System and classified as an asteroid as well as a dwarf planet.

1334. Several million years ago, a neighbor galaxy M32 plunged through the Andromeda Galaxy.

1335. Alan Shepard was the first American who went to space on the Mercury 3 on 5 May 1960. However, the spacecraft only traveled to 186 km (113 mi) above Earth before parachuting into the Atlantic Ocean.

1336. A massive bright sphere of scorching gas called plasma held together by its gravity is known as a star.

1337. NASA launched a Unity module aboard the Space Shuttle Endeavour, which was successfully attached to the Zarya module. The Unity module was fitted with all of the long-term human living requirements.

1338. The Andromeda and Milky Way galaxies may interact with the Triangulum Galaxy, and it may break it apart, turning it into an even larger elliptical galaxy!

1339. Meteor means 'suspended in air.'

1340. Martians have been written about in many novels. One of the most famous is H.G. Wells' 1898 novel The War of the Worlds, where Martians attempted to take over the world.

1341. On average, one meteoroid falls to Earth every year and burns up before hitting the ground.

1342. The Sun is 400 times the size of our moon, which is 400 times closer to Earth. For this reason, we can see solar eclipses on Earth.

1343. William Lassel found Triton, one of Neptune's moons, 17 days after Johann Gottfried Galle found Neptune.

1344. The pull of Earth's gravity is stronger at the poles, so a person standing at the Equator will be lighter and weigh 68 kg (150 lb) instead of 68.4 kg (150.8 lb) at the North Pole.

1345. The constellations that border the Cancer constellation are Lynx, Leo Minor, Leo, Hydra, Gemini, and Canis Minor.

1346. Some animals become confused when it gets dark in a total solar eclipse, and they prepare to go to sleep.

1347. Many comets are formed in two of the outermost regions of our solar system, the Oort Cloud and Kuiper Belts.

1348. In the year 150 AD, Ptolemy, a Greek astronomer, recorded 48 out of the 88 constellations that are officially recognized by the International Astronomical Union. The name of Ptolemy's book, where he recorded the constellations, is Almagest.

1349. The deepest known point on Earth, at 10,916 meters (35,814 feet) deep, is known as The Challenger Deep.

1350. The Hubble Space Telescope orbits Earth about 560 km (350 mi) above the planet.

1351. Saturn has the second shortest day of all the planets, lasting 10 hours and 34 minutes. Jupiter has the shortest.

1352. On 16 June 1963, Valentina Tereshkova became the first woman in space on Vostok 6. She was from the Soviet Union.

1353. There are different theories about the Big Bang. Some astronomers think that our universe was leftover from another universe. In contrast, others believe that we're just in the transition from something we know nothing about to one that we do.

1354. Martian dust storms can last for months and continuously change its surface.

1355. Jupiter's mass equals 2.5 times the combined mass of all other planets in our solar system.

1356. Both Buzz Aldrin and Neil Armstrong were initially quarantined for 21 days when they returned from the moon so that they didn't bring any disease back to Earth. At the time, astronomers didn't know there was no life on the moon.

1357. Just before the Battle of Hastings in 1066, Halley's Comet made an appearance.

1358. Supernovas can create shock waves strong enough to trigger new star formations.

1359. The Sun's mass occupies 99.86% of the mass of our solar system.
1360. The methane in Uranus' upper atmosphere absorbs the red light from the Sun and reflects the blue light into space. That's why it looks blue.
1361. On 11 December 2017, President Donald Trump signed the Space Policy Directive 1, which instructed NASA to send astronauts back to the moon and eventually to Mars.
1362. The Pinwheel Galaxy is about double the size of the Milky Way, but it looks faint as it is so far away.
1363. The tallest mountain in the solar system is on Mars.
1364. Mercury is slightly larger than our moon.
1365. Astronaut's vision is blurry in space as without gravity, body fluids in their body increase and puts pressure on their eyes.
1366. In a refracting telescope, the closer the lenses are, the blurrier the image.
1367. The Pinwheel Galaxy doesn't have a massive black hole at its center like most of the other galaxies. Instead, it has several smaller black holes.
1368. Uranus is four times bigger than Earth and the third largest planet in our solar system.
1369. Stars are mostly concentrated in the center of the Milky Way.
1370. Ceres may have been a planetary embryo that formed 4.57 billion years ago in the asteroid belt.
1371. Scientists believed the Milky Way was the center of our universe about 100 years ago.
1372. In the Mass Effect video game, Charon is a chunk of ice, not a moon.
1373. You can see about 2,000 stars without a microscope and about 50,000 using binoculars. If

you look through a 2" telescope, you can see 300,000 stars and much more with a 16" telescope.

1374. Scientists believe that a large planet-sized object hit Earth, breaking off rocks and debris, and these joined together to create our moon. This theory is known as the 'Giant Whack' or 'Giant Impact' theory.

1375. Two spacecraft have traveled to Mercury and a third on its way. Mariner 10 was a flyby mission that was launched in 1973 to Venus and Mercury. NASA's messenger was launched in 2004 and landed on Mercury in 2015. The last spacecraft, Bepi/Colombo, was launched in 2018 by ESA/JAXA and is due to arrive on Mercury in December 2025.

1376. The Hubble Space Telescope has observed a neutron star at the center of the Crab Nebula.

1377. Mars is often called the 'red planet' due to the iron minerals in its surface, causing it to look red.

1378. An astronomical satellite is one used for observations planets and galaxies.

1379. The lightest planet in our solar system is Saturn.

1380. About 50 active satellites are considered Medium Earth Orbit satellites, in orbit about 35,786 km (22,236 mi) above the Earth's surface.

1381. At 40,000 deg C (72,000 deg F), O-type stars are the hottest in the solar system.

1382. The Cancer constellation can be seen in the Northern hemisphere from late autumn to spring.

1383. The second fastest spinning planet is Saturn. Jupiter is the fastest.

1384. The moon represents a person's emotion and subconscious state in astrology. The sun is associated with fatherhood, while the moon is associated with motherhood.

1385. The Triangulum Galaxy is also known as Messier 33, NGC 598, and the Pinwheel Galaxy.

1386. An astronaut lost his wedding ring on a mission to the moon, but he found it again during a later spacewalk!

1387. The largest impact crater on Callisto, one of Jupiter's moons, is Valhalla, with a diameter of about 1900 km (1190 miles). The second-largest impact basin is Asgard, with a diameter of 1600 km (994 mi). Valhalla and Asgard are both locations where Odin and other gods ruled (Norse mythology).

1388. The "Karman Line," 100 km (62 miles) above sea level, is scientifically accepted as the edge of space. This means a total of 532 people from all over the world have reached outer space (as of June 2013).

1389. The temperature on the surface of Saturn is -139 deg C (-219 deg F).

1390. An Italian astronomer, Giovanni Schiaparelli, found strange lines on Mars in 1877. He called them Canali, which means 'channels' in Italian, but other nationalities misunderstood the translation and thought it meant 'canals.' Percival Lowell, an American astronomer, guessed incorrectly that the canals transported water from the Martian ice caps to the desert!

1391. Halley's Comet has an elliptical orbit around the sun.

1392. From our viewpoint on Earth, Jupiter seems to move slowly in the sky, taking months to move from one constellation to the next.

1393. The Pinwheel Galaxy is also known as Messier 101, M101, or NGC 5457.

1394. High tides, also known as spring tides, occur at full and new moon when the moon and sun line up with Earth. Neap tides occur when the sun and moon are at right angles to Earth.

1395. Charon is Pluto's largest moon.

1396. The only spacecraft to visit the Pluto system is the NASA spacecraft, New Horizons.

1397. A lenticular galaxy, for example, the Sombrero galaxy, resembles lenses and have features belonging to spiral and elliptical galaxies. They have a thin rotating disk of stars but no spiral arms. They are similar to elliptical galaxies as they have very little dust and matter, so they do not form new stars.

1398. Nearly all massive galaxies have a black hole at its center.

1399. NASA believes the astronauts on the first 1986 Challenger trip may have survived two minutes after it exploded. However, they would have then died when the spacecraft fell into the Atlantic Ocean at a speed of 321 kph (200 mph).

1400. Scientists think that there are about 2 trillion galaxies in the observable universe.

1401. Haumea is the third closest dwarf planet to the sun and the fourth largest dwarf planet.

1402. A shooting star is a meteor that burns up when it enters the Earth's atmosphere.

1403. The first space shuttle flew out on 12 April 1981.

1404. NASA's Mariner 9 was the first spacecraft to successfully orbit Mars in 1971, returning pictures of huge volcanoes and canyons, frozen underground ice and dried up rivers.

1405. Mercury has no moons or rings.

1406. The Kuiper Belt is a disc-shaped region that extends beyond Neptune. It contains hundreds of thousands of large icy objects bigger than 100 km (62 miles) in diameter and trillions of comets. Pluto is the most well-known object in the Kuiper Belt.

1407. The Spitzer Space Telescope is around the size of a car.

1408. About 200 asteroids, bigger than 100 km (62 mi) in diameter, have been identified in the Asteroid Belt. Hundreds of thousands more are smaller than this.

1409. A Dutch astronomer Christiaan Huygens was the first person to identify Saturn's rings, but he only saw one.

1410. Stars not only move around the center of the Milky Way, but they also move up and down.

1411. When Earth comes between the sun and the Moon, a lunar eclipse occurs. The Earth's shadow covers all or part of the Moon's surface.

1412. The result of a supernova is a black hole.

1413. The Hubble Space Telescope can capture colorful images of space objects such as dying stars or other galaxies.

1414. Earth rotates slower in March than it does in September.

1415. The furthest planet from the sun is Neptune.

1416. When new staff joins NASA, they are asked to identify inaccuracies in the movie Armageddon. More than 168 errors have been found.

1417. 99% of the mass of our solar system is made up of the Sun.

1418. A comet experiences heat when it nears the Sun, causing its ices to sublimate or sizzle. It may cause a small jet of material shooting out of the comet-like a mini geyser if the ice is close to its surface.

1419. Galileo used a refracting telescope that was shorter than 5 cm (2 in).

1420. If a rocket could travel at the speed of light, it would take it 100,000 years to travel across the Milky Way from one side to the other.

1421. Astronomers believe an explosion called Big Bang created the universe. The Milky Way formed not long after.

1422. Because there is no wind or rain on the moon, the astronaut's footprints from Apollo 11 will last for over 100 million years or more.

1423. A year on Earth is 365.2564 days which means we have an extra day in February every four years. This is called a leap year. Every leap year can be divided equally by four, so you can work out which years are leap years.

1424. Ceres has one thing that could sustain life that many other planets don't have: water.

Thank you for purchasing this book, we hope you enjoyed it!

What is your favorite planet?

Saturn is the Professor's favorite because of its massive rings that are made up of ice and rock! Did you know that the Professor created this beautiful cover … and he's only eight years old! The Professor loves all things animals, cars, science, food, and dinosaurs. And with his range of books, he hopes to inspire other kids to try new things, to learn more about the world, and to be curious!

We would really appreciate it if you have a couple of minutes to leave a 1-click review at bit.ly/space-review-p **- it would mean the world to us!**

* * *

To claim your free space coloring worksheet:
bit.ly/space-coloring

To see our other books:
amazon.com/author/professorsmart

To get in touch with us:
professorsmartpublishing@gmail.com